Modern electrical installation
for craft students

Modern electrical installation
for craft students
Volume 2

Third Edition

Brian Scaddan IEng., MIElec.IE

Heinemann Newnes

Heinemann Newnes
An imprint of Heinemann Professional Publishing Ltd
Halley Court, Jordan Hill, Oxford OX2 8EJ

OXFORD LONDON MELBOURNE AUCKLAND SINGAPORE
IBADAN NAIROBI GABORONE KINGSTON

First published by Butterworth & Co. (Publishers) Ltd 1983
Reprinted 1985
Third edition first published by Heinemann Professional Publishing 1990

© Heinemann Professional Publishing 1990

British Library Cataloguing in Publication Data
Scaddan, Brian
  Modern electrical installation for craft students. – 3rd ed.
  Vol. 2
  1. Buildings. Electric equipment. Installation
  I. Title
  621.319′24

ISBN 0 434 91848 2

Printed and bound by Biddles of Guildford

Contents

Preface

1	Inductors and Inductance	1
2	Capacitors and Capacitance	22
3	Inductance and Capacitance in Installation Work	33
4	Cells, Batteries and Transformers	45
5	Electrical Industries	70
6	Control and Earthing	110
7	Testing	121
8	Basic Electronics Technology	137

Miscellaneous Questions on Part I   151

Answers to Test Questions   169

Answers to Miscellaneous Questions on Part I   170

Index   172

To my father

Preface

This series of three books is designed to meet the syllabus requirements of Parts I and II of the City and Guilds 236 course in Electrical Installation Work, Volumes I and II covering Part I of the course, Volume III covering Part II.

Each book deals with craft theory, associated subjects, and a study of the electrical industries. The craft theory and associated subjects are treated, not as separate entities, but as a whole, encouraging the student to understand not just *how* but *why* electrical installations are designed and carried out in particular ways.

Wherever possible, calculations and examples have been devised to have values and wording appropriate to the work of the craft student, and the content of the volumes is comprehensive enough to relieve the student and staff of laborious note-taking. The object here is to encourage discussion and leave more time for demonstrations, individual investigation of the subject matter and classwork.

A great many useful diagrams, together with a simple method of approach ensure that students will have no difficulty in understanding either the English or the concepts.

The author wishes to express his thanks for all the help, encouragement and advice offered by his colleagues Mr R.V. Gorman, Mr C.A. Peach and Mr R. Bridger.

CHAPTER 1

Inductors and Inductance

Inductance: symbol, L; unit, henry (H)

Let us consider the effect of forming a coil from a length of wire, and connecting it to a d.c. source of supply. *Diagram 1* shows the distribution of the magnetic lines of force, or flux, produced by such a circuit.

If we wind the same coil onto an iron core, the lines of force tend to be confined to that core and the flux is much greater *(Diagram 2)*.

When a conductor is cut by magnetic lines of force, a current, and hence an e.m.f. is produced in that conductor. Consider then, what happens when the switch S is first closed *(Diagram 2)*.

As the current increases from zero to a maximum, the flux in the core also increases, and this growing magnetic field cuts the conductors of the coil, inducing an e.m.f. in them. This e.m.f., called the *back e.m.f.*, operates in the reverse direction to the supply voltage and opposes the change in the circuit current that is producing it. The effect of this opposition is to slow down the rate of change of current in the circuit.

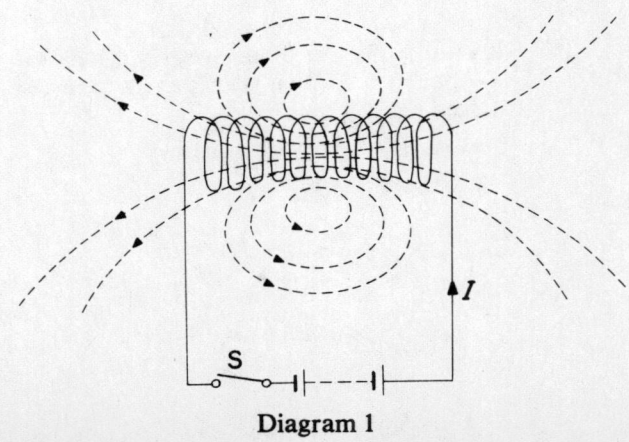

Diagram 1

2 Inductors and inductance

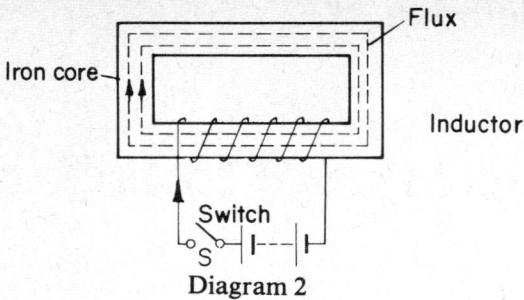

Diagram 2

When the switch S is opened, the current falls to zero and the magnetic field collapses. Again, lines of force cut the conductors of the coil inducing an e.m.f. in them. In this case, the e.m.f. appears across the switch contacts in the form of an arc.

Induced e.m.f. due to change in flux

The average value of the induced e.m.f. in a circuit such as the one shown in *Diagram 2* is dependent on the rate of change of flux, and the number of turns of the coil. Hence the average induced e.m.f.

$$E = -\frac{\Phi_2 - \Phi_1}{t} \times N \text{ volts}$$

The minus sign indicates that the e.m.f. is a back e.m.f., and is opposing the rate of change of current.

Example 1.1

The magnetic flux linking the 1800 turns of an electromagnet changes from 0.6 mWb to 0.5 mWb in 50 ms. Calculate the average value of the induced e.m.f. (E = induced e.m.f.)

$\Phi_2 = 0.6$ mWb; $\Phi_1 = 0.5$ mWb; $t = 50$ ms; $N = 1800$

$$E = -\frac{(\Phi_2 - \Phi_1)}{t} \times N$$

$$= -\frac{(0.6 - 0.5)}{50 \times 10^{-3}} \times 10^{-3} \times 1800$$

$$= -\frac{0.1 \times 1800}{50} = \frac{18}{5}$$

$$= -3.6 \text{ V}$$

Self inductance

Self inductance is the property of a coil in which a change of current and hence a change of flux, produces an e.m.f. in that coil.

The average induced e.m.f. in such a circuit is given by:

$$E = -\frac{L(I_2 - I_1)}{t} \quad \text{volts}$$

The inductance L can be calculated from:

$$L = \frac{N\Phi}{I}$$

where N = number of turns; Φ = flux in Wb; and I = current.

The unit of inductance

The unit of inductance is the *henry* (symbol H), and is defined as follows: *A circuit is said to possess an inductance of 1 henry, when an e.m.f. of 1 volt is induced in that circuit by a current changing at the rate of 1 ampere per second.*

Mutual inductance: symbol, M; unit, henry (H)

Let us consider the effect of winding two coils on the same iron core *(Diagram 3)*. This arrangement is called a *transformer*.

A change of current in coil 1 produces a change of flux which will link with coil 2, thus inducing an e.m.f. in that coil. These two coils are said to possess the property of *mutual inductance,* which is defined as: *A*

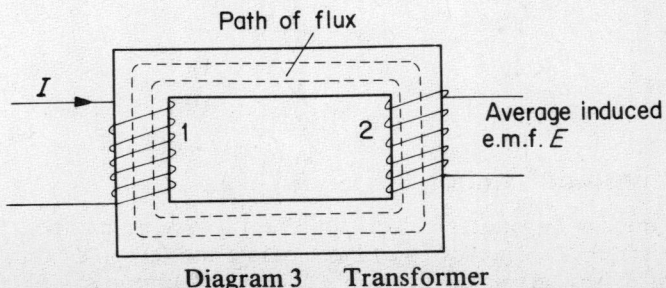

Diagram 3   Transformer

4 Inductors and inductance

mutual inductance of 1 henry exists between two coils, when a uniformly varying current of 1 ampere/second in one coil, produces an e.m.f. of 1 volt in the other coil.

If a change of current $(I_2 - I_1)$, in the first coil, induces an average e.m.f. E in the second coil, then:

$$E = -\frac{M(I_2-I_1)}{t} \text{ volts}$$

But as E can also be expressed as:

$$E = -\frac{(\Phi_2-\Phi_1)N}{t} \text{ volts}$$

Then:

$$\frac{M(I_2-I_1)}{t} = \frac{(\Phi_2 - \Phi_1)}{t} \times N$$

$$\therefore M = \frac{(\Phi_2 - \Phi_1)}{(I_2 - I_1)} \times N \text{ henrys}$$

Example 1.2

Two coils A and B have a mutual inductance of 0.5 H. If the current in coil A is varied from 6 A to 2 A, calculate the change in flux if coil B is wound with 500 turns.

$M = 0.5 \text{ H}; I_2 = 6 \text{ A}; I_1 = 2 \text{ A}; N = 500; (\Phi_2 - \Phi_1) = ?$

$$M = \frac{(\Phi_2 - \Phi_1)}{(I_2 - I_1)} \times N$$

$$(\Phi_2 - \Phi_1) = \frac{M \times (I_2 - I_1)}{N}$$

$$= \frac{0.5 \times (6-2)}{500}$$

$$= \frac{0.5 \times 4}{500}$$

$$= \frac{2}{500} = 4 \text{ mWb}$$

Time constant: symbol, T

When considering inductive circuits it is useful to represent the inductance and resistance of a coil as separate entities, on a circuit diagram. A typical inductive circuit is shown in *Diagram 4*.

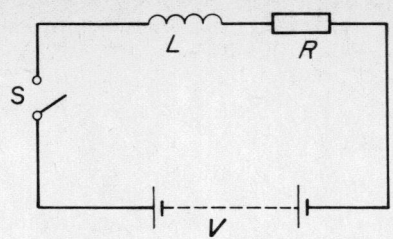

Diagram 4   Typical inductive circuit

When switch S is closed the current increases from zero to a steady maximum, given by:

$$I_{max} = \frac{V}{R} \text{ amperes}$$

Diagram 5 shows the growth of current in an inductive circuit. At any instant, say X, on the growth curve, if the rate of growth of current

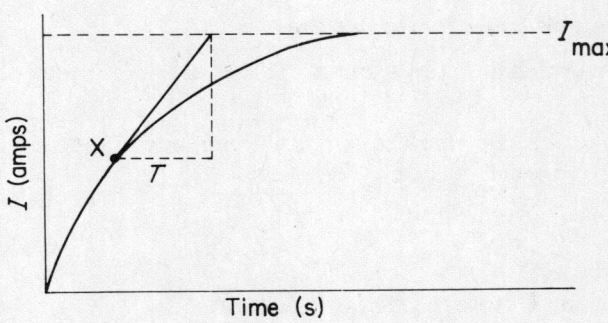

Diagram 5   Growth of current in an inductive circuit

at that instant is such that if it continued to increase at that rate, it would reach its maximum value in L/R seconds, then this period of time is called the time constant and is given by:

$$T = \frac{L}{R} \text{ seconds}$$

6 Inductors and inductance

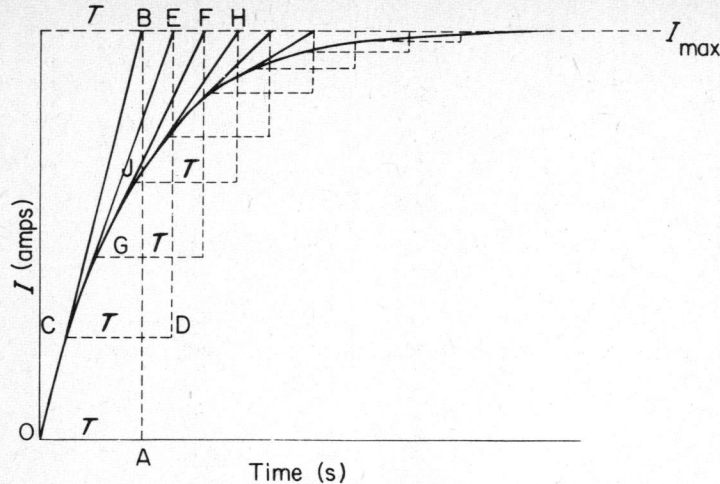

Diagram 6   Graphical representation of current growth curve

Graphical derivation of current growth curve

Construction *(Diagram 6)*

1. Select suitable scales for the two axes.
2. Draw the dotted line corresponding to the value of the maximum current (from $I = V/R$).
3. Along the time axis, mark off OA corresponding to the time constant T (from $T = L/R$).
4. Draw the perpendicular AB.
5. Join OB.
6. Select a point C close to O along OB.
7. Draw $CD = T$ horizontally.
8. Draw the perpendicular DE.
9. Join CE.
10. Repeat procedures 6 to 9 for the line CE, and continue in the same manner as shown in *Diagram 5*.
11. Join all the points O,C,G,J, etc. to form the growth curve.

The more points that are taken, the more accurate the final curve will be.

Example 1.3

A coil having a resistance of 25 Ω and an inductance of 2.5 H, is connected across a 50 V d.c. supply. Derive the curve of the current growth graphically.

$$I_{max} = \frac{V}{R} = \frac{50}{25} = 2 \text{ A; time constant, } T = \frac{L}{R} = \frac{2.5}{25} = 0.1 \text{ s;}$$

Scales: 10 cm = 1 A and 10 cm = 0.2 s

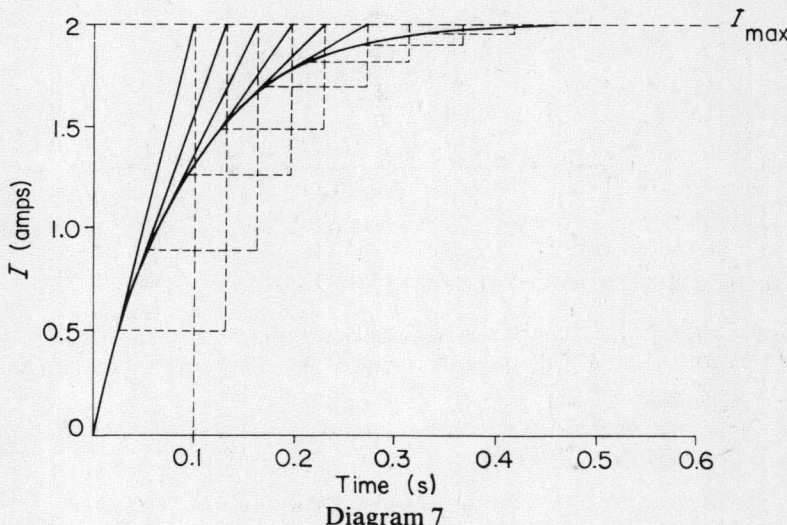

Diagram 7

Derivation of curve of current decay

The curve of current decay is constructed in the same manner as the growth curve, but in reverse as is shown in *Diagram 8*.

Energy stored in a magnetic field

As we have already seen, opening an inductive circuit produces an arc across the switch contacts. This arc is the dissipation of the magnetic energy which was stored in the coil; the value of this energy can be calculated from:

$$W = \frac{1}{2} \times L \times I^2 \text{ joules}$$

8 Inductors and inductance

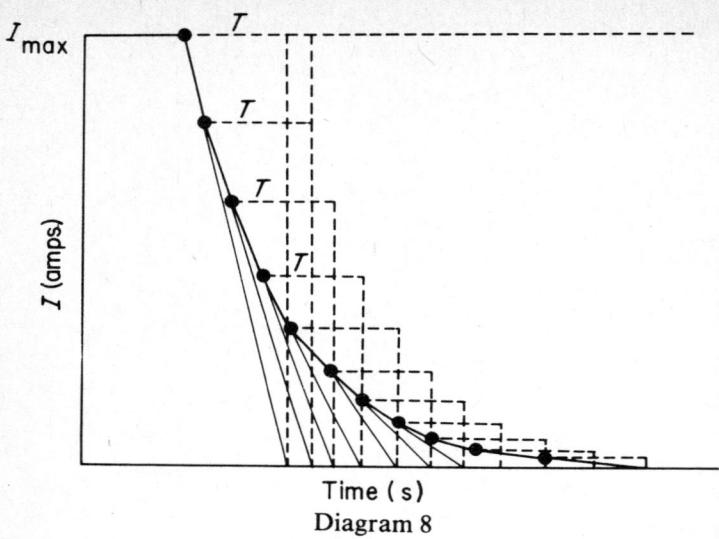

Diagram 8

Example 1.4

When carrying a current of 1.2 A, each field coil of a generator has an inductance of 2.5 H. Calculate the value of the energy stored in each coil.

W = energy stored; I = 1.2 A; L = 2.5 H

$$W = \tfrac{1}{2} \times L \times I^2$$

$$= \frac{2.5 \times 1.2 \times 1.2}{2}$$

$$= 1.8 \text{ J}$$

Inductance in a.c. circuits

Inductive reactance: symbol, — X_L; unit, ohm (Ω)

Let us now consider the effect of supplying an iron-cored coil of negligible resistance with an alternating current and voltage.
    In this instance the current, and therefore the magnetic field, is building up and collapsing (in the case of a 50 Hz supply) 50 times every second and hence a continual alternating e.m.f. is produced. As we have seen at the beginning of the chapter, the back e.m.f. opposes the change in circuit current which is producing that e.m.f. Therefore, under a.c. conditions the e.m.f. produces a *continual* opposition to the current

(much in the same way as resistance does in a resistive circuit). This opposition is called the *inductive reactance* (symbol X_L) and is measured in Ω. X_L is given by:

$$X_L = 2\pi f L \ \Omega$$

where: f = frequency in Hz and L = inductance in H.

Example 1.5

Calculate the inductive reactance of a coil of inductance 0.5 H when connected to a 50 Hz supply.

$X_L = ?; f = 50 \text{ Hz}; L = 0.5 \text{ H}$

$$\begin{aligned} X_L &= 2\pi f L \\ &= 2\pi \times 50 \times 0.5 \\ &= 2\pi \times 25 \\ &= 50\pi \\ &= 157.1 \ \Omega \end{aligned}$$

When an a.c. supply is given to a *pure* inductance the principles of Ohm's law may be applied, i.e. $V = I.X_L$

Example 1.6

Calculate the current taken by a coil of inductance 0.8 H when connected to a 100 V, 50 Hz supply.

$X_L = ?; V = 100 \text{ V}; f = 50 \text{ Hz}; L = 0.8 \text{ H and } I = ?$

In order to find the current, the formula $V = I \times X_L$ must be used, therefore the value of X_L must be calculated first.

$$\begin{aligned} X_L &= 2\pi f L \\ &= 2\pi \times 50 \times 0.8 \\ &= 80\pi \\ &= 251.36 \ \Omega \\ V &= I \times X_L \\ \therefore I &= \frac{V}{X_L} \\ &= \frac{100}{251.36} \\ &= 0.398 \text{ A} \end{aligned}$$

10 Inductors and inductance

Representation of current by a phasor diagram

In a purely resistive circuit *(Diagram 9a)* only the magnitude of the current is opposed by the resistance, and as the current and voltage alternate at the same time they are said to be *in phase*. Diagram 9b shows the waveforms of current and voltage in a resistive circuit.

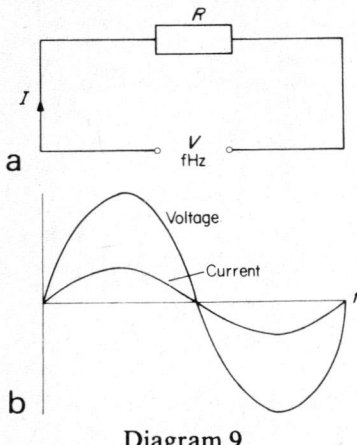

Diagram 9

In a purely inductive circuit *(Diagram 10a)* the rate of change of current is opposed by the reactance of the coil, and the effect of this opposition is to make the current lag behind the applied voltage or be *out of phase* by 90°. The waveforms of current and voltage in a purely inductive circuit are shown in *Diagram 10b*.

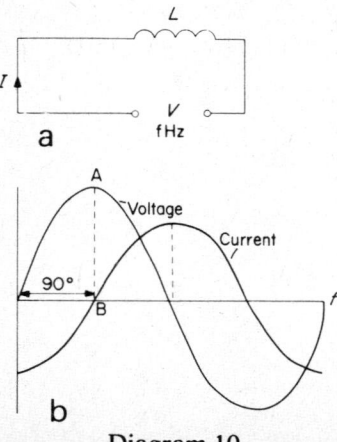

Diagram 10

The current lags the voltage by 90°, as V has reached its maximum at point A when current is at zero, point B.

We can represent this effect by means of *phasors* (scaled lines representing electrical quantities). *Diagram 11* shows the phasor representation of current and voltage in a purely resistive circuit.

Diagram 12 shows the phasor representation of current and voltage in a purely inductive circuit.

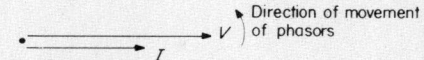

Diagram 11   Voltage and current in phase

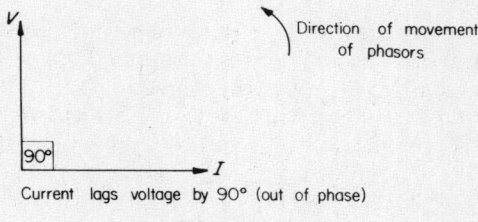

Diagram 12

Resistance and inductance in series (R-L circuits)

Consider a coil which has inductance and resistance as shown in *Diagram 13*.

It is clear that the applied voltage, V, comprises the voltage across L, V_L, and the voltage across R, V_R, the current remaining common. However, unlike purely resistive circuits, we cannot merely add V_L to

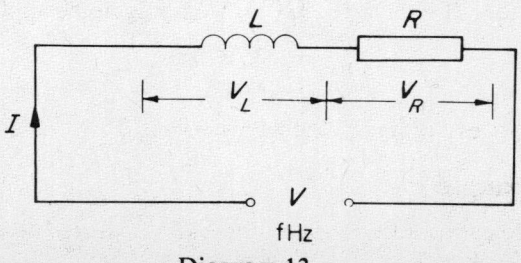

Diagram 13

12 Inductors and inductance

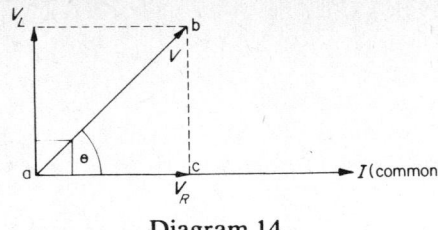

Diagram 14

V_R to obtain V. The reason for this is that in the inductive part of the circuit the common current I is out of phase with V_L and in the resistive part I is in phase with V_R. V_L and V_R can only, therefore, be added graphically (or by phasors) as in *Diagram 14*.

By construction we can see that V_R is in phase with I and V_L is 90° out of phase with I, and that the resultant is the applied voltage V. $\Theta°$ is the number of degrees that I lags behind the applied voltage V.

From *Diagram 14* it can be seen that the triangle abc is right-angled, having ab = V; bc = V_L and ac = V_R. Hence by Pythagoras' theorem:

$$ab = \sqrt{ac^2 + bc^2}$$

$$\therefore V = \sqrt{V_R^2 + V_L^2}$$

Note: This formula need not be remembered; it is simply a useful check after a value has been obtained with the aid of a phasor diagram.

Impedance: symbol, Z; unit, ohm (Ω)

It is clear that there are two separate oppositions to the flow of current in an R-L circuit, one due to resistance and the other due to reactance.

The combination of these oppositions is called the *impedance* of the circuit, its symbol is Z and it is measured in ohms.

Impedance may be defined as the *total* opposition offered by the components in that circuit. Ohm's law may once again be applied:

$$Z = \frac{V}{I}$$

where V is the applied voltage of the whole circuit.

Impedance triangle

From *Diagram 15* can be seen that a triangle can represent all the voltages in the circuit.

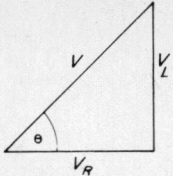
Diagram 15

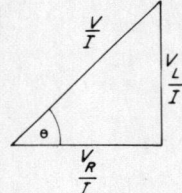
Diagram 16

If we now divide these voltages by the current *I* which is common to all components in the circuit the triangle can be shown as in *Diagram 16*.

$$\frac{V}{I} = Z \quad \frac{V_L}{I} = X_L \quad \frac{V_R}{I} = R$$

Therefore the triangle can be shown as in *Diagram 17*. This triangle is called the *impedance triangle*. Applying Pythagoras' theorem:

$$Z = \sqrt{R^2 + X_L^2}$$

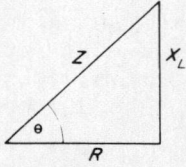
Diagram 17

Example 1.7

A choke coil has a resistance of 6 Ω and an inductance of 25.5 mH. If the current flowing in the coil is 10 A when connected to a 50 Hz supply *(Diagram 18)*, find the supply voltage *V*.

14 Inductors and inductance

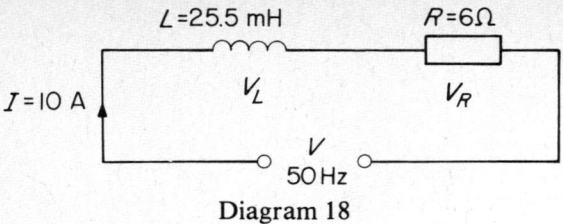

Diagram 18

In order to solve the problem by the use of phasors it is necessary to know the values of V_L and V_R. Hence:

$$V_R = I \times R = 10 \times 6 = 60 \text{ V}$$
$$V_L = I \times X_L$$
$$X_L = 2\pi fL$$
$$= 2\pi \times 50 \times 25.5 \times 10^{-3}$$
$$= 8 \, \Omega$$
$$\therefore V_L = I \times X_L$$
$$= 10 \times 8$$
$$= 80 \text{ V}$$

By phasors we can draw *Diagram 19* (scale 1 cm = 10 V).
By measurement $V = 100$ V.
Check by Pythagoras' theorem:

$$Z = \sqrt{R^2 + X_L^2}$$
$$= \sqrt{6^2 + 8^2}$$
$$= \sqrt{100}$$

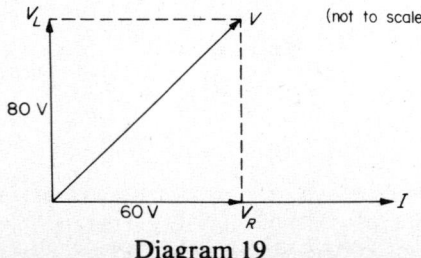

Diagram 19

$$= 10\,\Omega$$
$$V = I \times Z$$
$$\therefore V = 10 \times 10 = 100\text{ V}$$

Resistance and inductance in parallel

Again, unlike resistive circuits, currents in parallel branches of an *R-L* circuit cannot simply be added to find the total current. It will be seen from *Diagram 20a* that the common quantity in the circuit is the voltage. This is used as the reference phasor (as the current was in the series circuit).

The current in the resistive branch I_R is in phase with the applied voltage and the current in the inductive branch I_L lags the applied voltage by 90°. The resultant of these two currents is the supply current I *(Diagram 20b)*. The impedance Z is given by:

$$Z = \frac{V}{I}\,\Omega$$

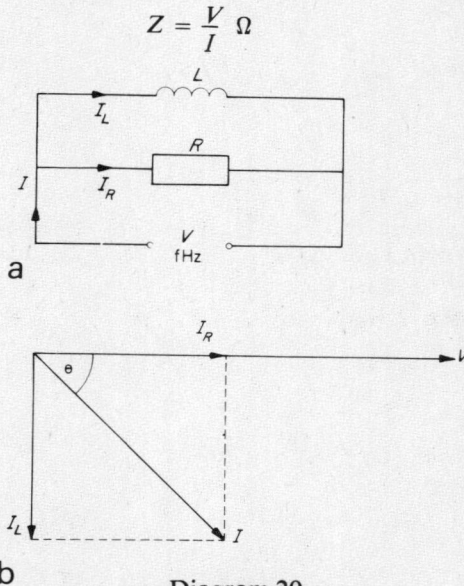

Diagram 20

Example 1.8

A non-resistive inductor of 51 mH is connected to a non-inductive resistor of 10 Ω across a 200 V, 50 Hz supply. What is the value of the supply current and impedance of the circuit? *(see Diagram 21).*

16 Inductors and inductance

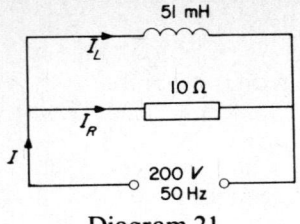

Diagram 21

In order to find I, the values of I_R and I_L must be found.

$$I_L = \frac{V}{X_L} \text{ and } I_R = \frac{V}{R}$$

$$\begin{aligned}X_L &= 2\pi f L \\ &= 2\pi \times 50 \times 51 \times 10^{-3} \\ &= 16\ \Omega\end{aligned}$$

$$\begin{aligned}\therefore I_L &= \frac{V}{X_L} = \frac{200}{16} \\ &= 12.5\text{ A}\end{aligned}$$

$$\begin{aligned}I_R &= \frac{V}{R} = \frac{200}{10} \\ &= 20\text{ A}\end{aligned}$$

By phasor diagram *(Diagram 22)* $I = 23.7$ A

$$\begin{aligned}Z &= \frac{V}{I} \\ &= \frac{200}{23.7} \\ &= 8.45\ \Omega\end{aligned}$$

```
         I_R = 20 A
   ┌─────────────────→ 200 V
   │
 θ │
   │
   │
   └ - - - - - - - ↘ I
 I_L = 12.5A
```

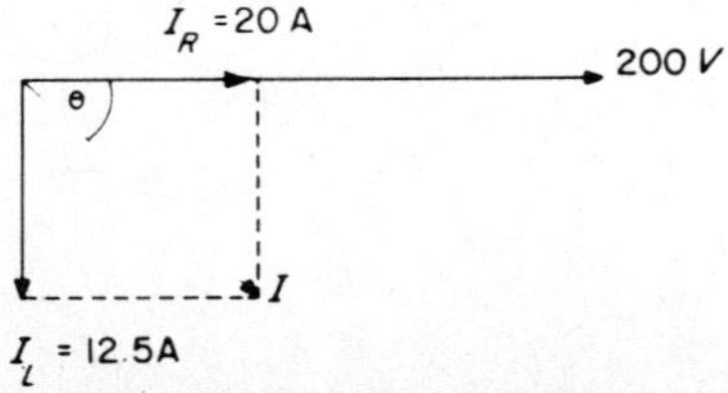

Diagram 22

Power

All the power (watts) in a circuit is dissipated in the circuit resistance. The purely inductive part of the circuit consumes *no* power, it only provides a magnetic field.

Consider the circuit shown in *Diagram 23*. We have already seen the phasor diagram of the voltage in the circuit and developed a voltage triangle from it. This time we *multiply* the voltage by the common current *(Diagram 24)*.

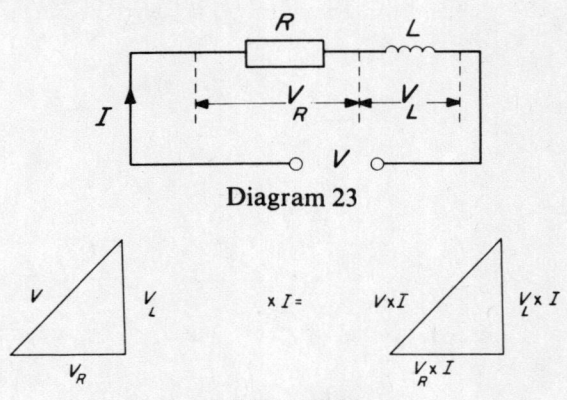

Diagram 23

Diagram 24

But voltage × current = power. Therefore the triangle becomes as shown in *Diagram 25*.

This is called the *power triangle* but as the inductive part of the circuit consumes no power, it is called the *wattless component* of power (VA,). The resistive part is called the *wattful component* or *true power* and the combination of the two is known as the *apparent power* (VA). The power triangle is usually shown in terms of kVA, kW and kVA, *(Diagram 26)*.

The relationship between the true power and the apparent power is very important.

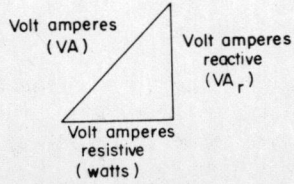

Diagram 25

18   Inductors and inductance

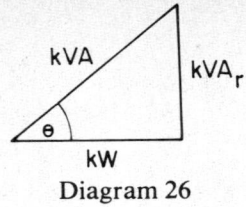

Diagram 26

Power factor

The ratio of the kW (true power) to the kVA (apparent power) is called the power factor (P.F.):

$$\text{P.F.} = \frac{\text{kW}}{\text{kVA}}$$

By trigonometry

$$\frac{\text{kW}}{\text{kVA}} = \cos \theta$$

$$\therefore \text{P.F.} = \cos \theta$$

As the voltage, impedance, and power triangles have the same angles, $\cos \theta$ is either:

$$\frac{V_R}{V} \text{ or } \frac{R}{Z} \text{ or } \frac{\text{kW}}{\text{kVA}}$$

all of which equal the power factor.

As the original triangle was formed from the phasor diagram *(Diagram 14)*. $\theta°$ is angle between the current and the supply voltage, therefore P.F. may be defined as: *the cosine of the angle of phase difference between the current and the applied voltage.*

Power factors in inductive circuits are termed *lagging* as the current lags the voltage.

When the true power equals the apparent power the P.F. of 1 is usually referred to as *unity*. Under these circumstances there would be no wattless power (kVA,) and the current taken by the circuit would be at a minimum. This is clearly an ideal situation.

The beer analogy

This is a useful way to explain the power factor. *Diagram 27* shows a pint beer glass with the main body of beer and the head. Although the glass is full, part of it is useless (remember this is only an analogy) and the true amount of beer is less than a pint. A ratio of true to apparent beer would indicate how much head there was. So, if this ratio (pint factor) were 1 or

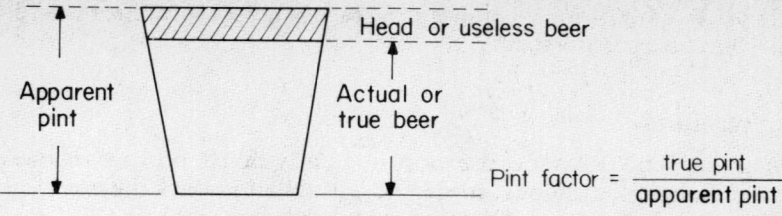

Diagram 27 Beer analogy

unity there would be no head, and a P.F. of 0.5 would mean half beer and half head. Clearly, it is better to have a P.F. close to unity.

All large plant (motors, transformers etc.) are rated in kVA, unlike most domestic appliances which are rated in kW. The reason for this is best explained by an example.

Example 1.9

If a heating appliance has a power rating of 1 kW at 240 V it will take a current of:

$$I = \frac{P}{V}$$
$$= \frac{1000}{240}$$
$$= 4.17 \text{ A}$$

But if a motor has a power rating of 1 kW at 240 V and the motor windings cause a P.F. of 0.6, then as:

$$\text{P.F.} = \frac{kW}{kVA}$$

$$kVA = \frac{kW}{\text{P.F.}}$$
$$= \frac{1}{0.6}$$
$$= 1.667 \text{ kVA}$$
$$= 1667 \text{ VA}$$

and since:

$$\text{current} = \frac{\text{voltamperes}}{\text{volts}}$$

$$I = \frac{1667}{240}$$
$$= 7 \text{ A}$$

20 Inductors and inductance

Had the cable supplying the motor been rated on the kW value it would clearly have been undersized.

Applications

Inductors, or chokes as they are more popularly called, are used in many areas of modern technology. In electrical installation work the main application is in fluorescent lighting, where the choke is open-circuited across the ends of the tube to cause it to strike. This effect is discussed further in Chapter 3. Motor windings are also inductances.

Questions on Chapter 1

1. If a coil has an e.m.f. of 6 V induced in it by a flux changing from zero to 36 mWb in 0.18 s, calculate the number of turns on the coil.

2. (a) What is self inductance?; (b) A relay coil of 300 turns produces a flux of 5 mWb when carrying a current of 1.5 A. Calculate the inductance of the coil.

3. An iron-cored coil having an inductance of 0.1 H and a resistance of 1.25 Ω, is connected to a 25 V d.c. supply. Calculate the circuit time constant and the maximum current. Draw to scale the curve of the current decay when the supply is switched off. What will be the value of current after 0.15 s?

4. (a) What is the effect of opening an inductive circuit? How can this effect be used?; (b) The energy stored in a coil is 2 J; if the inductance of the coil is 160 mH, calculate the coil current.

5. An inductor of negligible resistance has an inductance of 100 mH and an inductive reactance of 31.42 Ω when connected to an a.c. supply. Calculate the supply frequency.

6. A coil of inductance 1528 mH and negligible resistance takes a current of 0.5 A when connected to a 50 Hz supply. Calculate the value of the supply voltage.

7. An inductor has a resistance of 16 Ω and an inductive reactance of 12 Ω. If the current flowing in the circuit is 12 A, find, by means of a phasor diagram, the value of the supply voltage.

8. A pure inductance which has a reactance of 12 Ω is wired in parallel with a resistance of 8 Ω across a 240 V supply. Calculate the current in each component and determine the value of the supply current by means of a phasor diagram.

9. (a) Define the term *power factor*. (b) A circuit consists of a resistance and an inductance in series. The voltage across the resistance is 192 V, and the power factor is 0.8. Determine the value of the supply voltage and the voltage across the inductance.

10. An ammeter, a voltmeter and a wattmeter are to be connected into the circuit supplying a single-phase motor. Draw a diagram showing how these instruments would be connected. If the readings obtained were 240 V, 1.25 A and 180 W, calculate the power factor of the motor.

CHAPTER 2

Capacitors and Capacitance

Capacitors

A capacitor consists of two metal plates separated by an insulator, called a *dielectric;* the whole assembly is able to store electricity. This store is in the form of an excess of electrons on one plate and a deficiency on the other. In this state the capacitor is said to be charged. The charge is achieved by applying a voltage across the plates.

The type of capacitor commonly used in installation work is the electrolytic capacitor. This consists of plates of metal foil placed on either side of a waxed paper dielectric like a sandwich *(Diagram 28)*. It is manufactured in a long strip, rolled up and sealed into a metal container.

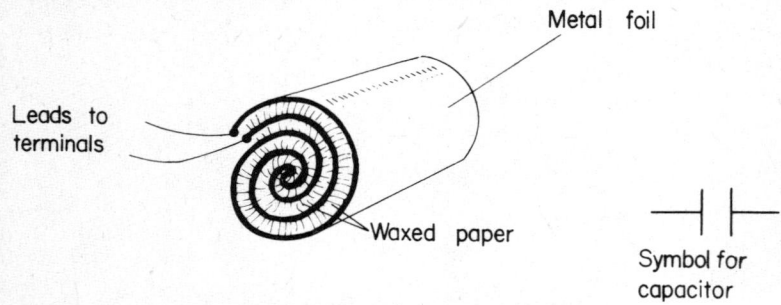

Diagram 28   Capacitor construction

Capacitance: symbol, C; unit, farad (F)

The unit of capacitance is the *farad* and may be defined as: *the capacitance of a capacitor which requires a potential difference of 1 volt to maintain a charge of 1 coulomb on that capacitor.*

Hence:

$$\text{charge} = \text{capacitance} \times \text{voltage}$$
$$Q\,(\text{C}) = C\,(\text{F}) \times V\,(\text{V})$$

Example 2.1

Calculate the charge on a 50 µF capacitor when it is connected across a 200 V d.c. supply:

$$Q = C \times V$$
$$= 50 \times 10^{-6} \times 200$$
$$= 0.01 \text{ C}$$

Dimensions of capacitors

If we take a simple parallel plate capacitor with an air dielectric, measure its capacitance, and then move the plates further apart, we find that the capacitance is smaller when measured a second time. We can therefore state that an increase in dielectric thickness (d) causes a decrease in capacitance. *Capacitance is inversely proportional to dielectric thickness.*

$$C \propto \frac{1}{d}$$

If, however, we were to keep the dielectric thickness constant and to vary the area of the plates (a), we would find that a change in plate area would cause a corresponding change in capacitance. The larger the plate area the larger the capacitance etc. *Capacitance is directly proportional to plate area.*

$$C \propto a$$

Combining these two effects we can see that:

$$C \propto \frac{a}{d}$$

Capacitors in series

Consider the effect of connecting three similar capacitors in series.

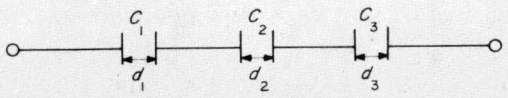

Diagram 29

24 Capacitors and capacitance

We know that:
$$C \propto \frac{1}{d}$$
$$\therefore d \propto \frac{1}{C}$$

So:
$$d_1 \propto \frac{1}{C_1} \quad d_2 \propto \frac{1}{C_2} \quad d_3 \propto \frac{1}{C_3}$$

If we were to combine all the dielectrics, we would have one capacitor of dielectric thickness d_{total} and capacitance C.

$$d_t \propto \frac{1}{C}$$

But:
$$d_t = d_1 + d_2 + d_3$$
$$= \frac{1}{C_1} + \frac{1}{C_2} + \frac{1}{C_3}$$

But:
$$d_t = \frac{1}{C}$$

$$\therefore \frac{1}{C} = \frac{1}{C_1} + \frac{1}{C_2} + \frac{1}{C_3} \ldots \text{etc.}$$

Just as the current is common to all parts of a series resistive circuit, so charge is common in a series capacitive circuit. Therefore:

$$Q = CV; \quad Q = C_1V_1; \quad Q = C_2V_2; \quad Q = C_3V_3 \text{ etc.}$$

Example 2.2

Three capacitors of 60 μF; 40 μF and 24 μF are connected in series across a 500 V d.c. supply. Calculate the total capacitance and the charge on each capacitor.

$$\frac{1}{C} = \frac{1}{C_1} + \frac{1}{C_2} + \frac{1}{C_3}$$

$$= \frac{1}{60} + \frac{1}{40} + \frac{1}{24}$$

$$\frac{1}{C} = \frac{1}{12}$$

$$\therefore C = 12 \, \mu F$$

Q is common to each capacitor

$$\therefore Q = CV$$
$$= 12 \times 10^{-6} \times 500$$
$$= 6 \times 10^{-3}$$
$$= 6 \text{ mC}$$

Capacitors in parallel

Let us arrange three similar capacitors in parallel *(see Diagram 30)*. We know that $C \propto a$. Therefore $C_1 \propto a_1$; $C_2 \propto a_2$ and $C_3 \propto a_3$

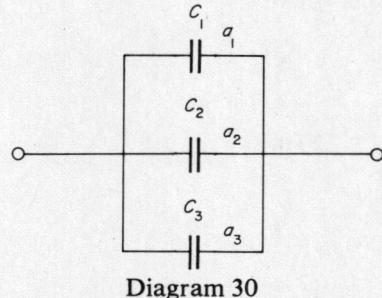

Diagram 30

As the plates connected to either side of the supply are common, we could replace the arrangement with one capacitor C of plate area a total.

$$\therefore a_T = a_1 + a_2 + a_3$$
$$\therefore a_T = C_1 + C_2 + C_3$$

But $C \propto a_T$

$$\therefore C = C_1 + C_2 + C_3$$

In this case it is the voltage that is common and the charge Q behaves like the current in a parallel resistive circuit. So:

$$Q = CV; Q_1 = C_1V; Q_2 = C_2V; Q_3 = C_3V$$

Example 2.3

Three capacitors of 60 μF, 40 μF and 24 μF are connected in parallel across a 500 V supply. Calculate the total capacitance, the total charge and the charge on each capacitor.

$$C = C_1 + C_2 + C_3$$
$$= 60 + 40 + 24$$
$$= 124\,\mu F$$

$$\text{Total charge } Q = C \times V$$
$$= 124 \times 10^{-6} \times 500$$
$$= 62\,mC$$

$$Q_1 = C_1 V$$
$$= 60 \times 10^{-6} \times 500$$
$$= 30\,mC$$

$$Q_2 = C_2 V$$
$$= 40 \times 10^{-6} \times 500$$
$$= 20\,mC$$

$$Q_3 = C_3 V$$
$$= 24 \times 10^{-6} \times 500$$
$$= 12\,mC$$

Energy stored in a capacitor

The amount of energy stored in a capacitor, is expressed in joules and is given by:

$$W = \tfrac{1}{2} C V^2$$

Capacitors in d.c. circuits

A capacitor connected across a d.c. supply is shown in *Diagram 31a*. The curves of the current and voltage in the circuit are shown in *Diagram 31b*.

As the capacitor begins to charge, its voltage increases until it is equal to the supply voltage. At the same time the charging current decreases. When the supply voltage and the capacitor voltage are equal the current in the circuit will be zero.

Diagram 32a shows the charged capacitor connected across a resistor. *Diagram 32b* shows the curves of the discharge voltage and current.

Capacitors and capacitance 27

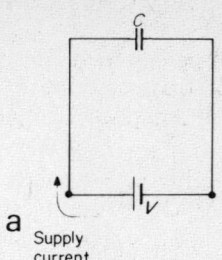
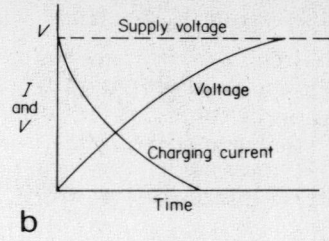

Diagram 31 Charging. (a) Capacitor connected across d.c. supply. (b) curves of discharge voltage and current.

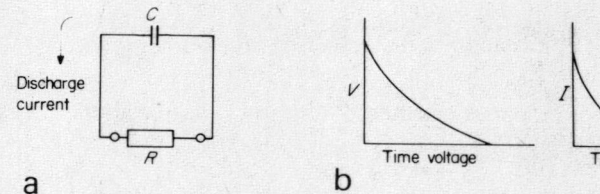

Diagram 32 Discharging. (a) Charged capacitor connected across a resistor; (b) curves of discharge voltage and current

Curves of current and voltage change

These curves are plotted in the same manner as those in inductive circuits (*see* Chapter 1).

Maximum charging or discharging current:

$$I = \frac{V}{R}$$

Time constant:

$$T = CR$$

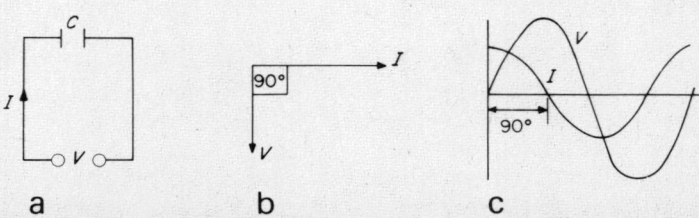

Diagram 33 (a) Circuit diagram; (b) phasor diagram; (c) waveforms

Capacitance in a.c. circuits

In an a.c. circuit a capacitance has the effect of opposing the voltage thus causing the circuit current to *lead*. In a purely capacitive circuit the current leads the voltage by 90°. The waveforms and phasors of such a circuit are shown in *Diagram 33*.

Capacitive reactance : symbol, X_C; unit, ohm (Ω)

The opposition offered by a capacitor in an a.c. circuit is called the *capacitive reactance* and is given by:

$$X_C = \frac{1}{2\pi f C}$$

where: X_C = capacitive reactance (Ω); f = frequency of supply (Hz); and C = capacitance (F).

As in the case of inductive reactance, Ohm's law may be applied, i.e.:

$$V = I \times X_C$$

Example 2.4

A purely capacitive circuit of 31.8 μF is connected to a 240 V, 50 Hz supply. Calculate the capacitive reactance and the circuit current.

$$\begin{aligned}
X_C &= \frac{1}{2\pi f C} \\
&= \frac{1}{2\pi \times 50 \times 31.8 \times 10^{-6}} \\
&= \frac{10^6}{100\pi \times 31.8} \\
&= 100\ \Omega \\
V &= I \times X_C \\
\therefore I &= \frac{V}{X_C} \\
&= \frac{240}{100} \\
&= 2.4\ \text{A}
\end{aligned}$$

Resistance and capacitance in series

Diagram 34 shows a capacitor in series with a resistor:
As the current leads the voltage across the capacitor and is in phase with the voltage across the resistor, the phasor diagram may be drawn as shown in *Diagram 35*.

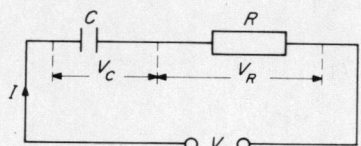

Diagram 34   Capacitor in series with resistor

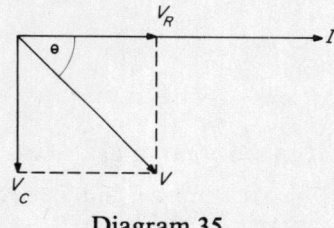

Diagram 35

Example 2.5

A capacitor of 159 μF is connected in series with a non-inductive resistor of 15 Ω across a 50 Hz supply. If the current drawn is 10 A calculate X_C, the voltage across each component and find by means of a phasor diagram the value of the supply voltage.

$$X_C = \frac{1}{2\pi fC}$$

$$= \frac{1}{2\pi \times 50 \times 159 \times 10^{-6}}$$

$$= \frac{10^6}{100\pi \times 159}$$

$$X_C = 20 \, \Omega$$

$$V_C = I \times X_C$$

$$= 10 \times 20$$

Capacitors and capacitance

$$V_C = 200 \text{ V}$$
$$V_R = I \times R$$
$$= 10 \times 15$$
$$V_R = 150 \text{ V}$$

By measurement *(Diagram 36)* V will be found to be 250 V.

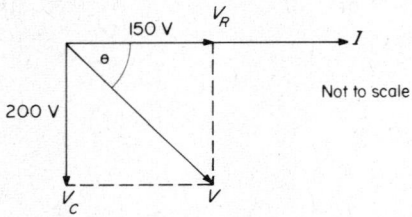

Diagram 36   Phasor diagram

Resistance and capacitance in parallel

In this case it is the voltage that is common and the currents that are added by phasors *(see Diagrams 37 and 38)*.

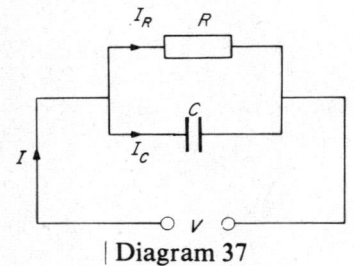

Diagram 37

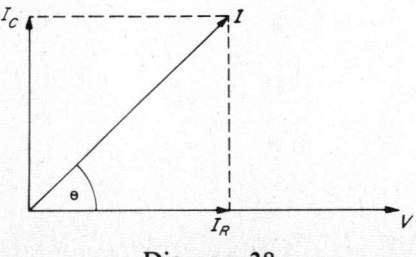

Diagram 38

Working voltage

Every capacitor has the value of its working voltage marked on it. Beyond this value the dielectric would break down and the capacitor would be useless.

Applications of capacitors

Capacitors are used extensively in electrical engineering. In the field of installation work, they are mainly used for motor starting, power factor correction and radio interference suppression and to minimize the stroboscopic effects in fluorescent lighting circuits. Their use in power factor correction and fluorescent lighting is dealt with in the next chapter.

Questions on Chapter 2

1. A capacitor has a value of 73 μF and is connected across a 100 V supply. Calculate the charge on the capacitor.

2. Three capacitors of 20 μF, 80 μF and 16 μF are connected in series across a 240 V supply. Calculate the charge.

3. A variable capacitor has a capacitance of 100 μF when the distance between the plates is 1 mm. What will the capacitance be if the plates are adjusted to be 10 mm apart?

4. Four capacitors of 10 μF, 20μF, 15μF, and 5 μF are connected in parallel across a 240 V supply. Calculate the total capacitance, the total charge, and the charge across the 20 μF capacitor.

5. A capacitor has a value of 150 μF and a plate area of 60 cm². What would be the plate area of a similar type of capacitor of 200 μF and the same dielectric thickness?

6. A parallel plate capacitor has a value of 636 μF. At what value of supply frequency would the reactance be 5 Ω?

7. A resistor of 6 Ω is connected in series with a capacitor of 398 μF, the current drawn being 24 A. Calculate the voltage across each component and find the supply voltage by means of a phasor diagram ($f = 50$ Hz).

32 Capacitors and capacitance

8. Draw the waveforms and phasor diagram for an a.c. current and voltage in a purely capacitive circuit.

9. A 127.3 µF capacitor is connected in parallel with a 50 Ω resistor across a 240 V supply. Calculate the current taken by each component. Determine the value of the supply current ($f = 50$ Hz).

10. Explain with the aid of a sketch the construction of an electrolytic capacitor. What is meant by the term *working voltage?*

CHAPTER 3

Inductance and Capacitance in Installation Work

In Chapters 1 and 2 we have discussed separately the effects of inductance and capacitance in an a.c. circuit. Here we consider how these effects may be applied, and in some cases, combined in a.c. circuits.

The magnetic effect of an inductor has many uses. It is, however, in equipment such as motors and fluorescent lighting that its effect on the power factor is substantial enough to cause concern, and make it necessary to improve the power factor.

Power factor improvement

Consider the diagram of a resistive and inductive circuit *(Diagram 39)*. The phasor diagram for this circuit is shown in *Diagram 40*.

We can show the supply current and voltage as in *Diagram 41*. If we redraw this phasor diagram so that the voltage is drawn horizontally it becomes as shown in *Diagram 42*.

If we now connect a variable capacitor across the supply terminals of the original load we have the result shown in *Diagram 43*. It is a parallel circuit and the voltage is common to both branches. We can therefore draw a current phasor diagram *(Diagram 44)*. As *V* is common, we can combine both diagrams *(Diagram 45)*.

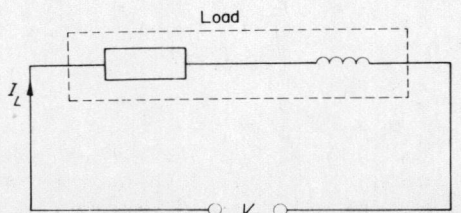

Diagram 39   Resistive and inductive circuit

34   *Inductance and capacitance in installation work*

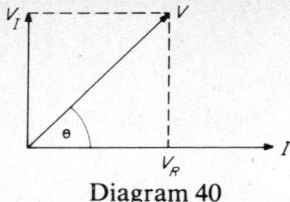
Diagram 40

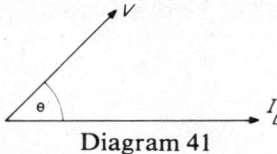
Diagram 41

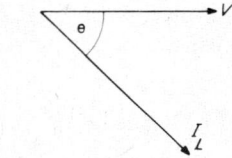
Diagram 42   Phasor diagram of load

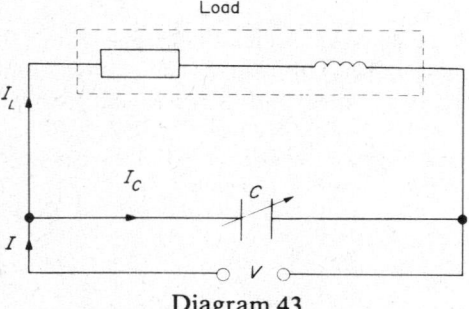
Diagram 43

The phasor of I_L and I_C *(Diagram 45)* is clearly I, which is of course the supply current. Note that it is smaller than the load current I_L and that the angle between I and V is smaller than that between I_L and V. The closer an angle is to zero the nearer its cosine is to unity. Therefore, the addition of the capacitor has improved the power factor of the system.

Inductance and capacitance in installation work

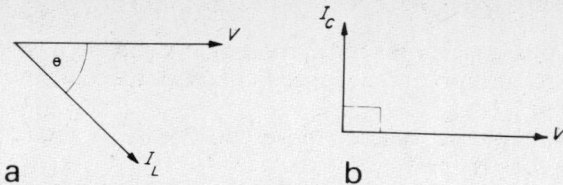

Diagram 44 (a) Phasor diagram for load, (b) phasor diagram for capacitor

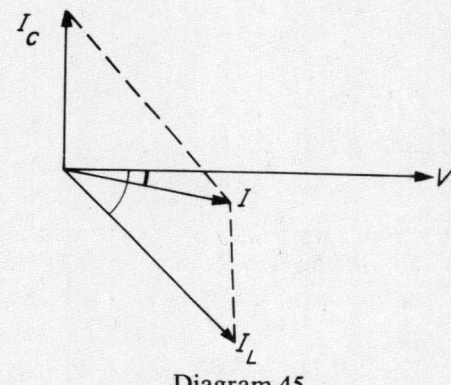

Diagram 45

The actual current taken by the load does not change, it is the total supply current that decreases. This means that smaller supply cables may be used. With industrial loads, the supply authority's transformer and switchgear as well as their cable may be reduced in size.

In order to encourage power factor (P.F.) improvement the supply authorities make a higher charge to consumers who do not correct or improve their P.F. to a suitable level (usually about 0.95 lagging). It is not usual to improve the P.F. much beyond this point as the cost of providing extra capacitance required to gain a small decrease in current is uneconomic.

Capacitors are the most popular method of improving the P.F., although synchronous motors are used occasionally (this is discussed in greater detail in Volume 3).

P.F. improvement capacitors may be fitted to individual plant or in banks connected to the supply intake terminals. The first method is more popular as the banked type needs automatic variation as plant is switched on and off.

Example 3.1

A 240 V single-phase motor takes a current of 8 A and has a power factor of 0.7 lagging. A capacitor is connected in parallel with the motor and takes a current of 3.1 A.

Draw a scaled phasor diagram of the currents in the circuit and find the value of the supply current, and the new P.F.

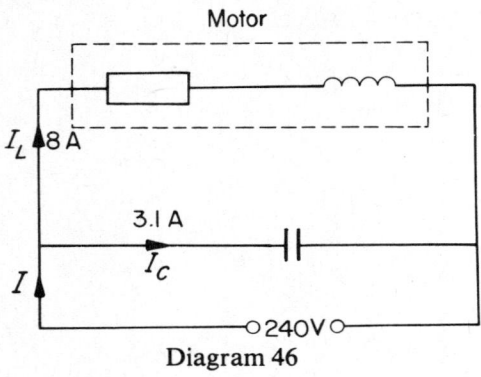

Diagram 46

From *Diagram 46*

$$I_t = 8 \text{ A}$$
$$\text{P.F.} = 0.7$$
$$\text{as P.F.} = \cos \theta$$
$$\text{then } \cos \theta = 0.7$$
$$\theta = 45.5°$$

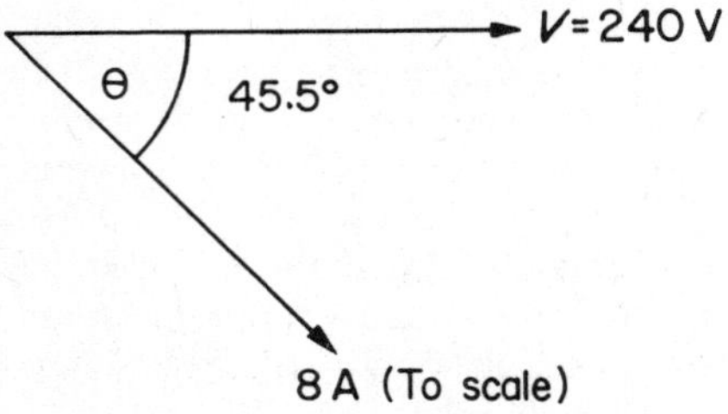

Diagram 47

Therefore the diagram for the motor is *Diagram 47*.

$$I_C = 3.1 \text{ A}$$
I_C leads V by 90°

Inductance and capacitance in installation work

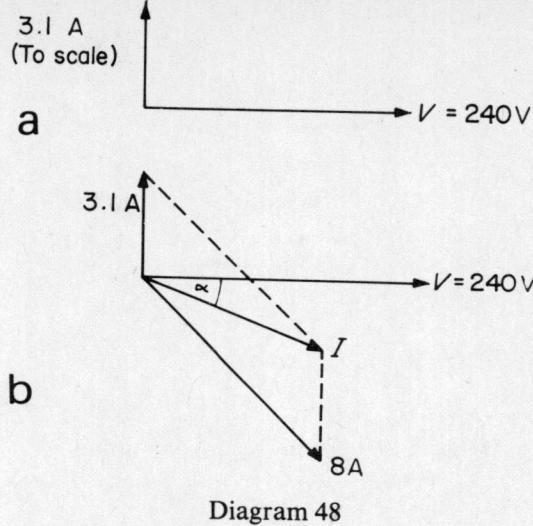

Diagram 48

Therefore the phasor diagram for the capacitor is *Diagram 48a*. Combining both phasor diagrams, we have *Diagram 48b*. By measurement:

$$I = 6.2 \text{ A}$$
$$= 24.5°$$
$$\therefore \text{P.F.} = \cos \alpha$$
$$\text{P.F.} = \cos 24.5°$$
$$= 0.91 \text{ lagging}$$

Inductors and capacitors in fluorescent lighting

The common fluorescent lamp consists of a glass tube filled with mercury vapour at a low pressure, and a little argon gas. The inside of the tube is coated with a fluorescent powder. There is an element at each end of the tube whose terminations are taken outside the tube to metal pins.

Theory of operation

If a high voltage is applied across a quantity of gas, electrons are liberated from some of the atoms of the gas and collide with other gas atoms. When this happens energy is released in the form of light. This process is called *ionization*.

38 *Inductance and capacitance in installation work*

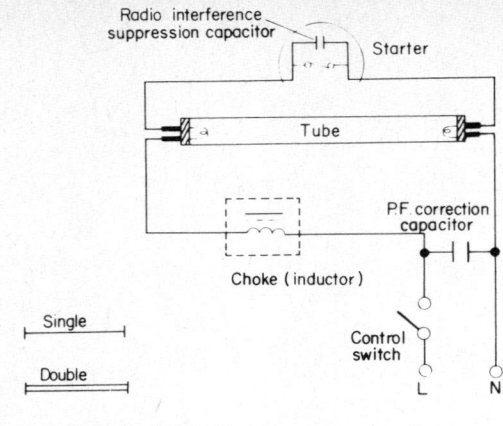

Symbols for fluorescent lamps

Diagram 49   Fluorescent light circuit

Practical operation

Diagram 49 shows the circuit for a typical fluorescent light unit.

When the supply is switched on, the circuit is completed via: the choke, first lamp element, starter switch, second lamp element and the neutral. The elements, which are coated in oxide, become warm and the oxide coating emits some electrons and the gas ionizes at the ends of the tube (this helps the main ionization process). The starter contacts (usually of the bimetallic type) separate, owing to the current passing through them, and the choke is open circuited.

As we have seen, breaking an inductive circuit causes high voltages to appear across the breaking contacts, and energy is released in the form of an arc. In this case, however, there is an easier way for the energy to dissipate — via the gas, and the high voltage appears across the ends of the tube.

When the gas is fully ionized, the choke limits the current to a pre-determined value, and the light emitted, which is mostly ultra-violet, is made visible by the fluorescent powder coating.

Note: Other types of discharge lamp and starter details are discussed more fully in Volume 3.

The radio interference suppression capacitor is usually located in the starter. The P.F. correction capacitor is part of the control circuitry common to all fluorescent lighting installations.

Example 3.2

The P.F. correction capacitor in a 240 V, 50 Hz fluorescent light unit has broken down and needs replacing. A test on the unit shows that without

Inductance and capacitance in installation work

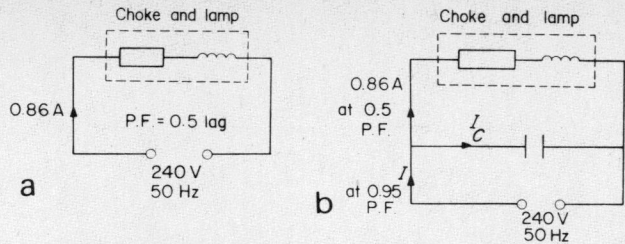

Diagram 50   (a) No capacitor, (b) with capacitor

the capacitor, the supply current is 0.86 A at a P.F. of 0.5 lagging. The values quoted on the original capacitor have faded and the only other information is that the working P.F. of the unit should be 0.95. Determine the value of the capacitor needed *(Diagram 50.)*

First the phasor diagram for the unit without a capacitor is drawn *(Diagram 51)*. P.F. = 0.5 and cos θ = 0.5. Therefore: θ = 60°.

The phasor diagram of the unit showing the supply current at working P.F. is drawn *(Diagram 52)*.

$$\text{P.F.} = 0.95$$
$$\cos \alpha = 0.95$$
$$\alpha = 18.2°$$

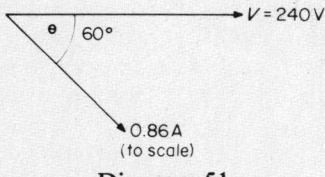

Diagram 51

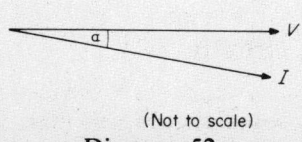

(Not to scale)
Diagram 52

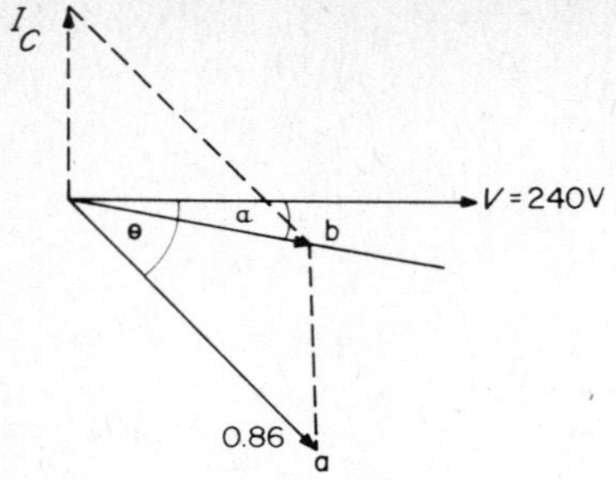

40 Inductance and capacitance in installation work

Diagram 53

Finally the combined phasor diagram can be drawn *(Diagram 53)*. The value of the capacitor current required to raise the P.F. to 0.95 lagging must be I_c, which is the same distance (ab). By measurement:

$$I_c = 0.6 \text{ A}$$

$$\therefore X_c = \frac{V}{I_c}$$

$$= \frac{240}{0.6}$$

$$= 400 \text{ }\Omega$$

But

$$X_c = \frac{1}{2\pi f C}$$

$$\therefore C = \frac{1}{2\pi f X_c}$$

$$= \frac{1}{2\pi \times 50 \times 400}$$

$$= 8\mu\text{F}$$

Note: By measurement, the supply current is 0.45 A.

Rating of fluorescent circuits

Fluorescent tubes are rated in watts, but as we have seen, the circuit of which the lamp is part, is inductive, and even after improvement, has a lagging P.F.

Inductance and capacitance in installation work

We know from Chapter 1 that plant is not rated in watts, but in voltamps (VA). The I.E.E. Regulations recommend that, if no other information is available, *the lamp wattage may be multiplied by 1.8 in order to determine the VA rating*. For example the VA rating of a fluorescent unit with an 80 W tube is $1.8 \times 80 = 144$ VA. Hence, when supplied at 240 V, the current taken would be:

$$I = \frac{VA}{V}$$

$$= \frac{144}{240}$$

$$= 0.6 \text{ A}$$

When a fluorescent lamp is switched off, the choke is again open circuited; this time the voltage appears across the switch contacts. The I.E.E. Regulations recommend that, unless the switch is specially designed to break an inductive circuit, *it must have a rating of not less than twice the total steady current it is required to carry*. For example: if a fluorescent light unit draws a current of 1 A, then the switch controlling it must have a rating of at least 2 A.

Example 3.3
A consumer has a work area that he wishes to illuminate with single 65 W, 240 V fluorescent fittings. The existing lighting points, which are to be removed are controlled by a single-gang 5 A switch. This switch is to remain. How many 60 W fittings may be installed?

$$\text{VA rating of fitting} = 1.8 \times 65$$
$$= 117 \text{ VA}$$
$$\therefore \text{Current rating of fitting} = \frac{117}{240}$$
$$= 0.4875 \text{ A}$$
$$\text{The 5 A switch must only carry } \frac{5}{2} = 2.5 \text{ A}$$
$$\therefore \text{Number of fittings permitted} = \frac{2.5}{0.4875}$$
$$= 5.13$$
$$= 5 \text{ fittings}$$

Stroboscopic effect

While a fluorescent lamp is in operation the light may flicker. Under some circumstances this may make it appear that rotating machinery has

42 Inductance and capacitance in installation work

slowed down or even stopped. This is called a *stroboscopic effect*. This is an undesirable state of affairs which is usually remedied by one of the two following methods.

Balancing the lighting load (three-phase)

If a large lighting load is installed in a three-phase installation where there is some rotating machinery, the stroboscopic effect may be overcome by connecting alternate groups of lamps to a different phase. This also has the advantage of balancing the lighting load *(Diagram 54)*.

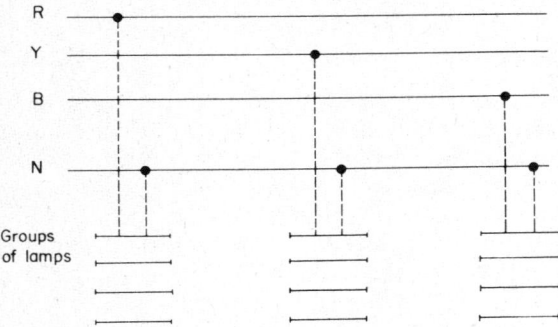

Diagram 54   Lighting load

The lead–lag circuit

In this method a capacitor is wired in series with every alternate lamp in a group. The value of the capacitor is such that the lamp unit it is fitted to has an overall *leading P.F.* This means that any pair of lamps have a lagging and a leading P.F. This has the effect of cancelling out the resultant flicker, in the same way as two equal but *opposing* forces cancel each other out. See *Diagram 55*.

Example 3.4

Two 240 V fluorescent lamp units A and B are arranged to overcome stroboscopic effects. Unit A has a series capacitor fitted and takes a current of 0.4 A at a P.F. of 0.985 leading. Unit B takes a current of 0.53 A at 0.5 P.F. lagging *(Diagram 56)*. Draw a scaled phasor diagram showing these two currents and from it determine the total current and the overall P.F. Ignore P.F. improvement.

Choose a suitable scale *(Diagram 57)*.

$$\cos \alpha = 0.985$$
$$\therefore \alpha = 10°$$

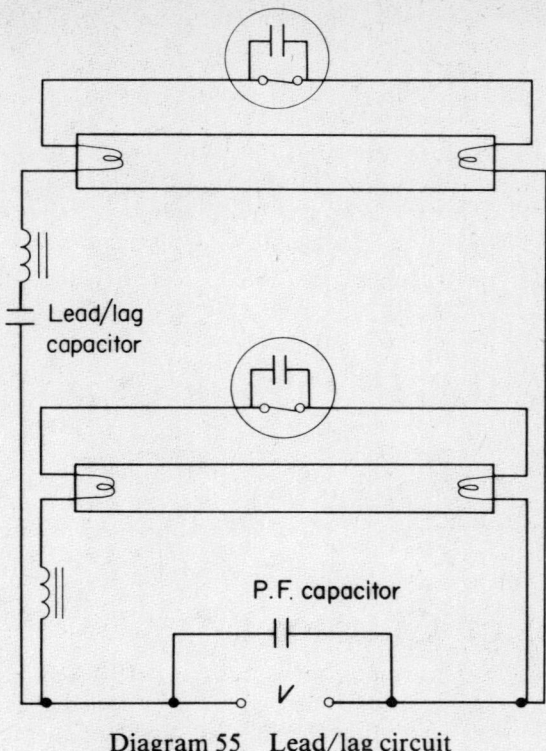

Diagram 55   Lead/lag circuit

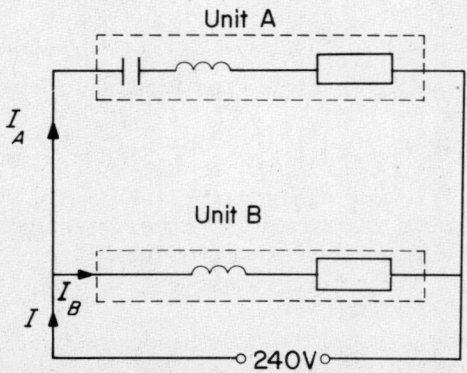

Diagram 56 $I_A = 0.4$ A at 0.985 lead. $I_B = 0.53$ A at 0.5 lag

44 Inductance and capacitance in installation work

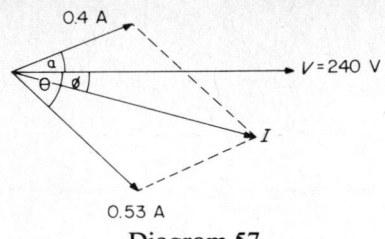

Diagram 57

$$\cos \theta = 0.5$$
$$\therefore \theta = 60°$$

By measurement:

$$I = 0.77$$
$$\phi = 30°$$
$$\therefore \text{P.F.} = \cos \phi$$
$$= 0.866$$

Questions on Chapter 3

1. A 240 V single-phase motor takes a current of 10 A and has a working power factor of 0.5 lagging. Draw a scaled phasor diagram and from it determine the value of capacitor current required to improve the P.F. to 0.9 lagging. Calculate the value of the capacitor.

2. What are the requirements of the I.E.E. Regulations with regard to (a) rating of fluorescent lamp units, and (b) current rating of ordinary switches controlling fluorescent lighting circuits?

3. With the aid of a diagram explain the principles of operation of a fluorescent light unit.

4. What is meant by the *stroboscopic effect,* and how can it be minimized?

5. The following data relate to two 240 V fluorescent lighting units arranged to minimize stroboscopic effect:
   Unit 1 — 0.75 A at 0.96 P.F. leading.
   Unit 2 — 0.8 A at 0.6 P.F. lagging.
   By using a scaled phasor diagram determine the value of the total current taken by the two units.

CHAPTER 4

Cells, Batteries and Transformers

General background

In 1789, an Italian professor of botany, Luigi Galvani (1737–98), noticed by chance that freshly skinned frogs' legs twitched when touched by two dissimilar metals. He called this effect *animal electricity*.

It was, however, another Italian, Alessandro Volta (1745–1827), a professor of physics, who showed that the electric current which produced the muscular spasm was not due to the animal limb itself, but to the moisture in it. In 1799 he developed a simple battery comprising copper and zinc discs separated by brine-soaked cloth. This type of assembly is known as a *voltaic pile*. From this primitive beginning have come the cells and batteries we use today. The materials used may be more refined, but the basic concept has remained unchanged.

The primary cell

If two dissimilar metals are immersed in an acid or salt solution, known as an *electrolyte*, an e.m.f. will be produced; this assembly is known as a *cell*. The e.m.f. may be used to supply a load, but will only do so for a limited time, as the chemical qualities of the electrolyte deteriorate with use. The chemicals have to be renewed to render the cell useful again.

The most common forms of primary cells in use are the Leclanché wet cell and the dry cell.

Diagram 58 illustrates the component parts of these two types of cell.

The depolarizing agent is used to remove hydrogen bubbles from around the carbon rod. These bubbles, which are formed during the chemical action, impair the performance of the cell.

Applications

The dry cell has an obvious advantage over the wet cell because it is portable and so is commonly used for appliances such as torches, door bells, etc.

46 Cells, batteries and transformers

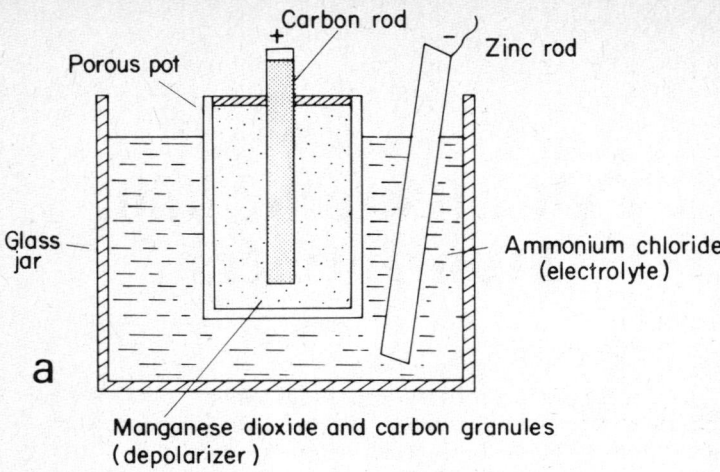

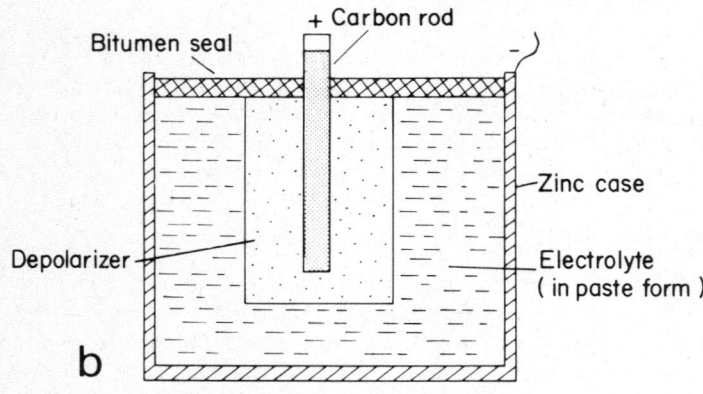

Diagram 58 (a) Wet cell; (b) dry cell

The wet cell, although almost obsolete, is used in larger bell and indicator circuits and for railway signalling.

The secondary cell

Unlike the primary cell, the secondary cell can be used again after it has discharged all its electrical energy. It can be recharged by *supplying* it with electrical energy. This reverses the chemical process which took place during discharge.

There are two types of secondary cell, the lead-acid and the alkaline cell.

Cells, batteries and transformers

The lead-acid cell

This cell consists of positive and negative lead electrodes, and an electrolyte of *dilute sulphuric acid* all placed in an acid-resistant container.

The electrodes are made of several plates, the positive and negative being insulated from one another by separators of insulating material such as wood or ebonite.

The construction of the plates is of considerable importance and is discussed below.

Formed plates

Repeated charging and discharging of a cell under manufacture causes the plates to be covered in lead compounds, the negative plate with *spongy lead* and the positive plate with *lead peroxide,* these being important to the chemical action of the cell. This process is, however, both expensive and time-consuming and for smaller types of cell, pasted plates are more popular.

Pasted plates

These plates are manufactured in the form of a grid, into which a compound of *sulphuric acid and red lead* is pressed; only a short initial charge is needed for the cell to be ready for use. These plates, however, disintegrate more easily than the formed type.

A combination of formed and pasted plates is used in large-capacity cells; the positive plate is formed and the negative plate is pasted.

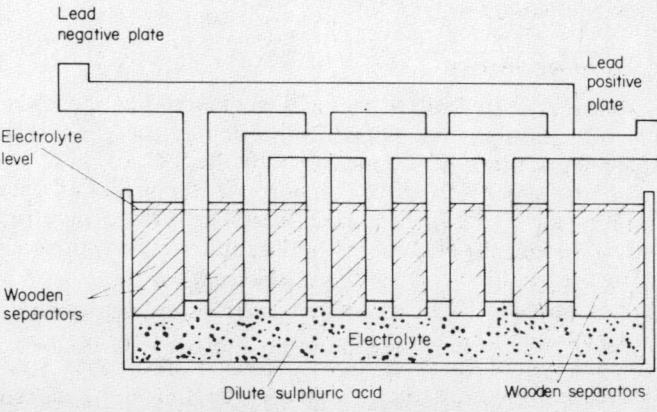

Diagram 59   Lead-acid cell

When current is drawn from the cell the active chemicals on the positive plate expand and the plate tends to distort, especially under heavy loads. Some measure of protection against this distortion or buckling, is achieved by arranging for each positive plate to be adjacent to two negative plates *(Diagram 59)*.

Action of lead-acid cell

When an external load is connected to the terminals of the cell, electrical energy is delivered to the load. During this *discharge* period, a chemical reaction between the plates and the electrolyte takes place and a layer of *lead sulphate* is deposited on the plates. However, this process successively weakens the electrolyte until the cell is unable to deliver any more electrical energy.

If a d.c. supply is then connected to the cell terminals, and a current is passed through it, the *lead sulphate* is converted back into *sulphuric acid* and restores the cell to its original condition. This process is known as *charging*.

Care and maintenance of lead-acid cells

Provided that a lead-acid cell is maintained regularly and is cared for, it should last for an indefinite period of time. A *weekly* check on its condition is to be recommended.

Electrolyte level

The level of the electrolyte should never be allowed to fall below the tops of the plates. Any loss of electrolyte due to evaporation may be made up by the addition of distilled water.

Specific gravity of electrolyte

As a cell discharges, the electrolyte becomes weaker and its specific gravity (S.G.) falls, until the cell can no longer deliver energy. The state of charge of a cell can therefore be measured by the S.G. of the electrolyte. A hydrometer is used for this purpose. It consists of a glass syringe containing a weighted, graduated float. The syringe has a rubber nozzle for insertion into the electrolyte, and a rubber bulb at the upper end for sucking the liquid into the syringe *(Diagram 60)*.

The nozzle is inserted in the electrolyte and a sample is drawn up into the syringe by squeezing the rubber bulb. The level of the liquid in relation to the position of the float, gives a direct reading of the S.G. of the electrolyte. The higher the float, the higher the S.G. *(see Diagram 60)*; the lower the float, the lower the S.G.

Cells, batteries and transformers 49

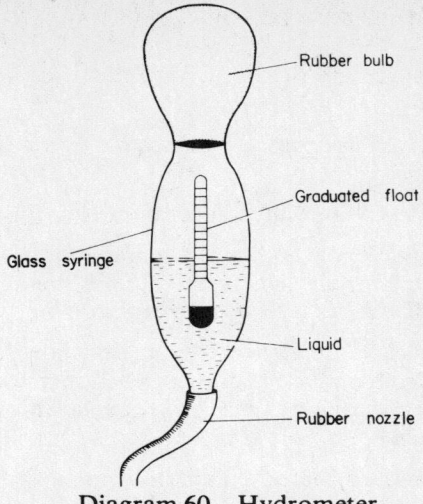

Diagram 60 Hydrometer

The following table indicates typical values of the specific gravity in relation to charge:

S.G.	Percentage of charge
1.28	100
1.25	75
1.22	50
1.19	25
1.16	Fully discharged

A record of results obtained should be kept for each cell.

Note: Cells should not be allowed to fall below a S.G. of 1.18.

Terminal voltage

A check on the no-load terminal voltage should be made with a high-resistance voltmeter at the end of a discharge period. This reading should not be below 1.85 V. A fully charged cell should indicate about 2.2 V.

Plate colour

An indication of the state of charge is the colour of the plates. In a healthy cell the positive plate is chocolate brown and the negative plate is slate-grey.

General maintenance procedures

1. When preparing an electrolyte, *always* add acid to water, *never* water to acid.
2. Ensure that any maintenance is carried out in a well-ventilated area.
3. Do not permit the use of any naked flame near the cells.
4. Cells which are to be taken out of commission for any time should be fully charged, the electrolyte left in, and a periodic charge given to keep the cell healthy until it is needed again.
5. Never leave a cell in an uncharged state, as a layer of whitish *sulphate* will form on the plates, which will increase the internal resistance and reduce the capacity of the cell. This process is known as *sulphation of the plates*.
6. Terminals should be coated with petroleum jelly to prevent corrosion.

Applications of lead-acid cells

The most common use for this type of cell is the car battery (a battery is a group of cells). Other applications include standby supplies, alarm and control circuits.

The nickel-alkaline cell

There are two types of alkaline cell, the nickel-iron and the nickel-cadmium.

The nickel-iron cell

Here the positive plate is made of *nickel hydroxide,* the negative plate of *iron oxide* and the electrolyte is *potassium hydroxide*.

The nickel-cadmium cell

In this cell both the positive plate and the electrolyte are the same as for the nickel-iron cell; however, the negative plate is *cadmium* mixed with a small amount of iron.

The active chemicals in the plates of alkaline cells are enclosed in thin nickel-steel grids insulated from one another by ebonite rods. The whole assembly is housed in a welded steel container.

Care and maintenance of the nickel-alkaline cell

Unlike the lead acid cell the nickel-alkaline needs minimal attention.
The open type only needs periodic topping up to compensate for the electrolyte lost by evaporation. The totally enclosed type needs no maintenance.

Applications

It has limited use owing to its cost and is mainly used in situations where a robust construction is needed, i.e. marine work.

Capacity of a cell

If a cell delivers, say, 10 A for a period of 10 hours, it is said to have a capacity of 100 ampere-hours (A h) at the 10 hour rate; taking any more than 10 A will discharge the cell in less than 10 hours.

Efficiency of a cell

The efficiency of any system is the ratio of the output to the input. The efficiency of cells is given in two forms:

The ampere-hour efficiency % $= \dfrac{\text{discharge amperes} \times \text{time} \times 100}{\text{charging amperes} \times \text{time}}$

The watt-hour efficiency % $= \dfrac{\text{discharge } VA \times \text{time} \times 100}{\text{charging } VA \times \text{time}}$

Comparison of cell characteristics

	Lead–acid	Alkaline
S.G. charged	1.28	1.2
S.G. discharge	1.18	1.2
P. d. charged	2.1 V	1.3 V
P.d. discharge	1.85 V	1.0 V

Ah capacity

The A h obtainable from the alkaline cell at the higher discharge rates, i.e. 2 and 4 hours, is much greater than that of the lead acid cell *(Diagram 61)*. This is because the S.G. of the electrolyte does not change during discharge.

Charge and discharge

The dips in the charging curves *(Diagram 62)* are due to the reduction of the charging current, to prevent overheating of the cell.

	Advantages	Disadvantages
Lead–acid	Inexpensive High discharge voltage Uses plentiful materials	Fragile Self-discharges when not in use Requires regular maintenance
Alkaline	Very robust Retains its charge when not in use Needs little or no maintenance	Very expensive Low discharge voltage

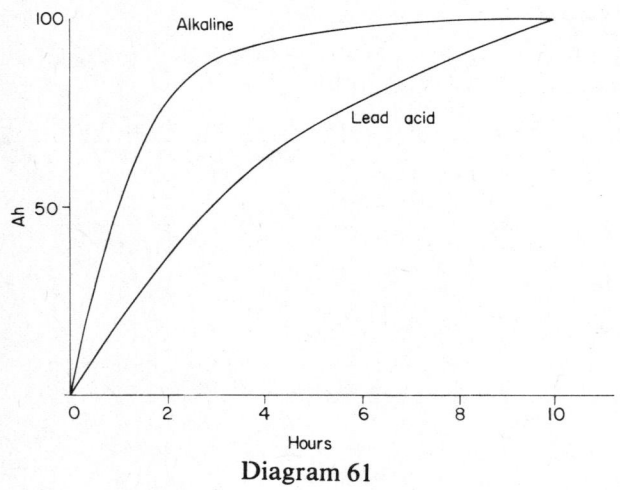

Diagram 61

Cell and battery circuits

E.m.f. of cell (E, volts)

This is the maximum force available (measured in volts) in a cell to produce current flow.

Internal resistance

When current flows through the cell there is some resistance to its flow (less than 1 Ω in a good cell) and hence a voltage drop across it.

Cells, batteries and transformers 53

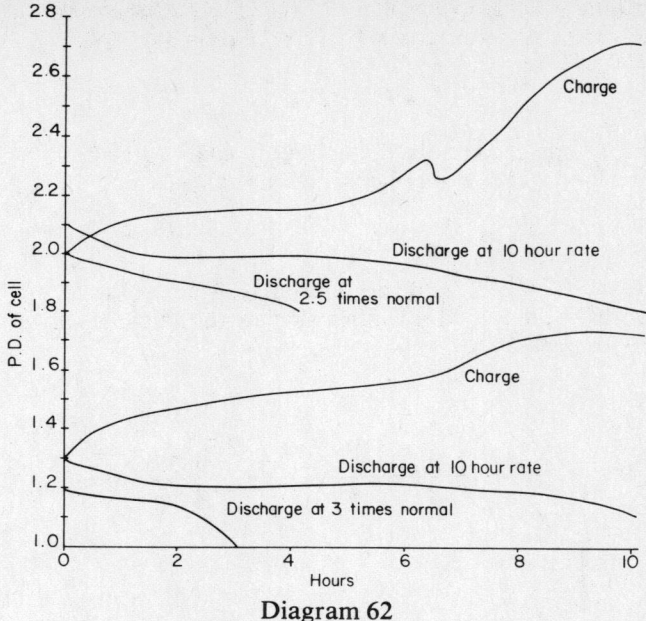

Diagram 62

P.d. of a cell or terminal voltage (*V*, volts)

This is the voltage measured at the terminals of a cell, and is less than the cell e.m.f. owing to the voltage drop across the internal resistance of the cell *(Diagram 63)*. E = e.m.f. of cell; r = internal resistance of cell; R = resistance of load; I = circuit current and V = terminal voltage or p.d. across load.

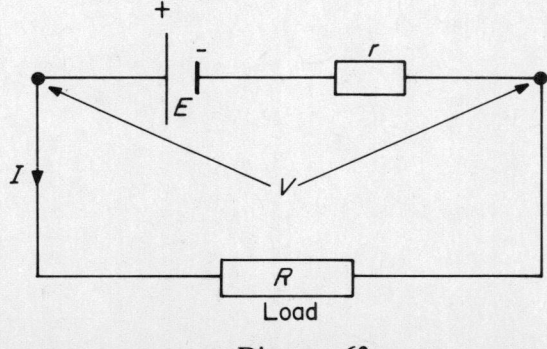

Diagram 63

54 Cells, batteries and transformers

The terminal voltage *(V)* available across the load is clearly the e.m.f. *(E)* less the volt drop across the internal resistance *(r)*.

$$V = E - (I \times r)$$

This is the same principle as the voltage available across the terminals of a load which is supplied by a long cable (Volume 1, Chapter 3).

Example 4.1

If a cell with an e.m.f. of 2 V and an internal resistance of 0.2 Ω is connected across a 0.8 Ω load resistor, calculate the current that will flow *(see Diagram 64)*.

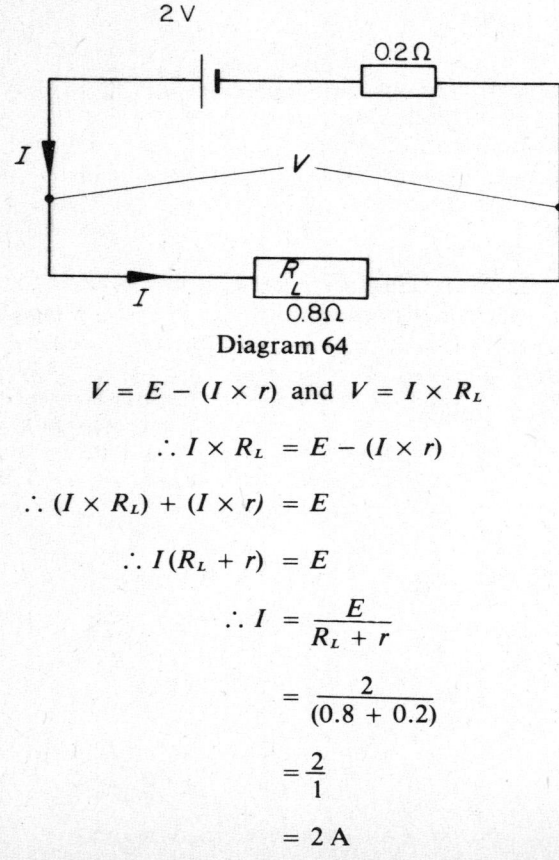

Diagram 64

$$V = E - (I \times r) \text{ and } V = I \times R_L$$

$$\therefore I \times R_L = E - (I \times r)$$

$$\therefore (I \times R_L) + (I \times r) = E$$

$$\therefore I(R_L + r) = E$$

$$\therefore I = \frac{E}{R_L + r}$$

$$= \frac{2}{(0.8 + 0.2)}$$

$$= \frac{2}{1}$$

$$= 2 \text{ A}$$

Diagram 65   Cells in series

Cells in series

If a high p.d. is required, then cells are connected in series and internal resistances are added *(Diagram 65)*.

Cells in parallel

For cells in parallel the p.d. is the same as that for one cell, but as the internal resistances are added in parallel their resultant internal resistance is less than for one cell and heavier currents can be drawn.

Battery charging

Cells and batteries are charged by connecting them to a controlled d.c. source. This source may be obtained in several ways: (1) rectified a.c.; (2) motor generator set; (3) rotary converter and (4) d.c. mains supply.

The most commonly used method is rectified a.c. and there are two ways in which this system is used: (1) the constant-voltage method, and (2) the constant-current method.

Constant-voltage charging

In this method the d.c. charging voltage is kept constant at a value just above that of the final value of the battery e.m.f. The charging current is initially high, decreasing as the e.m.f. of the battery approaches that of the supply *(Diagram 66)*.

Constant-current charging

In constant-current charging the current is kept constant by varying the d.c. input voltage as the battery e.m.f. increases *(Diagram 67)*.

The more popular method, for everyday use, is the constant-voltage method.

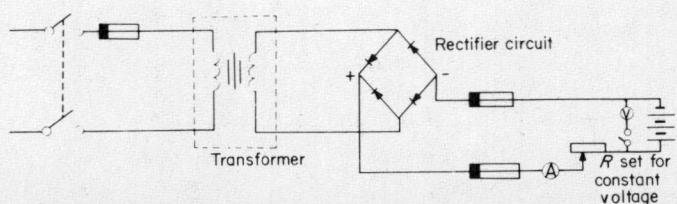

Diagram 66   Constant-voltage charging

56 Cells, batteries and transformers

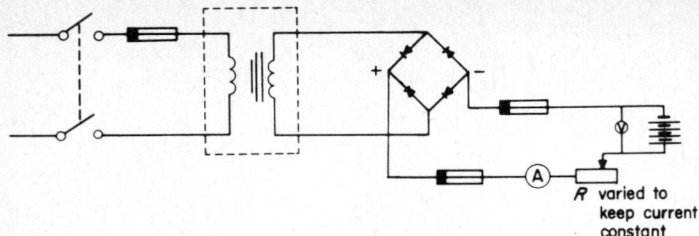

Diagram 67 Constant-current charging

Cells, batteries and their associated charging equipment are frequently used in installation work such as indicator and call systems in hospitals and hotels, fire alarm and burglar alarm systems, and emergency lighting installations. Whatever the application, somewhere in the circuitry of either the main or the charging system, transformers and rectifiers are used.

Transformers

As we have seen in Chapter 1, two coils that are wound on the same iron core have the property of mutual inductance, because a change in flux, and hence in e.m.f., in one coil produces, via the iron core, a corresponding change in the other coil.

If we take the same arrangement and apply an alternating voltage to one coil, it will induce an alternating e.m.f. in the other coil; this is called the *transformer effect.* The coil or winding to which the supply is connected is called the *primary* and the winding from which the induced voltage is taken is called the *secondary (Diagram 68).*

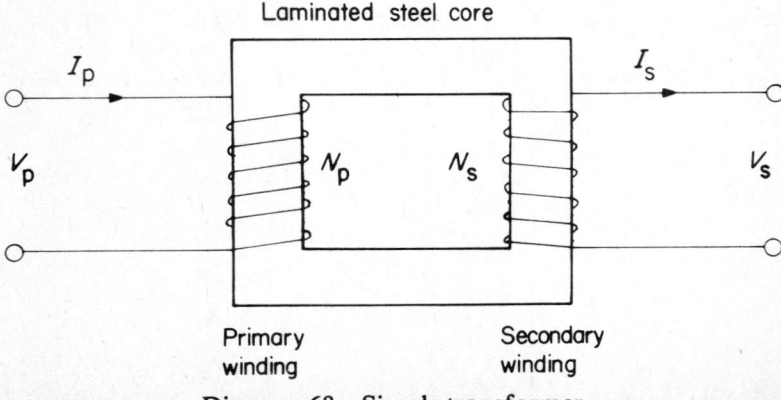

Diagram 68 Simple transformer

Cells, batteries and transformers

The relationship between the voltage, current and number of turns for each winding is as follows:

$$\frac{V_p}{V_s} = \frac{N_p}{N_s} = \frac{I_s}{I_p}$$

where: V_p = primary voltage; I_p = primary current; N_p = primary turns; V_s = secondary voltage; I_s = secondary current and N_s = secondary turns.

Transformers which have a greater secondary voltage are called *step-up* transformers, while those with a smaller secondary voltage are called *step-down* transformers.

Example 4.2

A single-phase step-down transformer has 796 turns on the primary and 365 turns on the secondary winding. If the primary voltage is 240 V calculate the secondary voltage. Also calculate the secondary current if the primary current is 10 A.

$$\frac{V_p}{V_s} = \frac{N_p}{N_s}$$

$$\therefore V_s = \frac{V_p \times N_s}{N_p}$$

$$= \frac{240 \times 365}{796}$$

$$= 110 \text{ V}$$

also $\frac{V_p}{V_s} = \frac{I_s}{I_p}$

$$\therefore I_s = \frac{I_p \times V_p}{V_s}$$

$$= \frac{10 \times 240}{110}$$

$$= 21.82 \text{ A}$$

Note the larger secondary current. The secondary winding would need to have a larger conductor size than the primary winding to carry this current. If the transformer were of the step-up type the secondary current would be smaller.

Types of transformer

Double-wound

This type is constructed as shown in *Diagram 68*. Two electrically separate coils are wound onto a common silicon steel core.

The core is laminated to lessen the effects of eddy currents and silicon steel is preferred, as there are few losses due to hysteresis. These losses are dealt with later in this section.

The double-wound transformer is the commonest form of transformer, and has a wide range of applications.

Auto-transformer

In this type of transformer a single coil is wound onto a steel core, the primary and secondary windings being part of one winding *(Diagram 69)*.

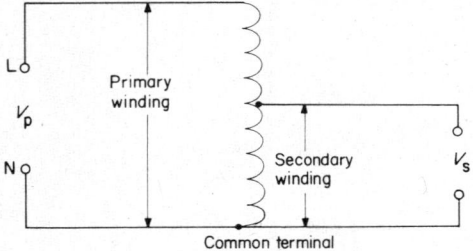

Diagram 69   Auto-transformer

The main use of this type of transformer is in the grid system. When 400 000 V (400 kV) has to be transformed (stepped-down) to 132 kV, huge transformers are required. Auto-transformers are used mainly because, as there is only one winding, a great saving in copper and hence expense is achieved.

The main disadvantage in using auto-transformers for applications such as bells or train sets, etc. is that the primary and secondary windings are *not* electrically separate and a short-circuit on the upper part of the winding (*see Diagram 69*) would result in the whole of the primary voltage appearing across the secondary terminals.

The same ratio applies between the voltages, currents and number of turns.

The current transformer

The action of this transformer is the same as those previously discussed. It is a step-up (voltage) transformer and is used extensively for taking

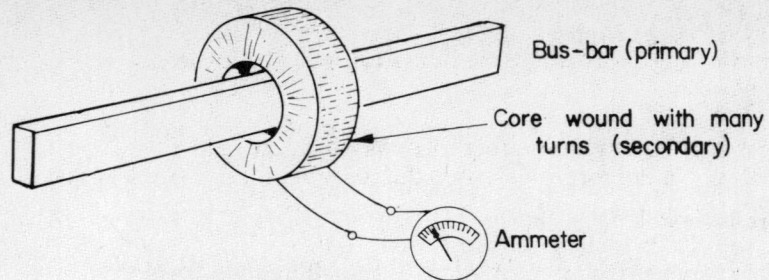

Diagram 70 Bar primary transformer

measurements. The most common form is the bar primary type *(Diagram 70)*.

It is clearly impracticable to construct an instrument to measure currents as high as say 200–300 A, so a current transformer is used to step down the secondary current to a value which can be measured on a standard instrument.

Example 4.3

A current of 300 A flowing in a bus-bar needs to be measured. The ammeter available has a maximum rating of 0.5 A. How many turns on the secondary of a current transformer would be required to measure the primary current? where $N_p = 1$ (single bus-bar); $I_p = 300$ A and $I_s = 0.5$ A (instrument rating).

$$\frac{I_s}{I_p} = \frac{N_p}{N_s}$$

$$N_s = \frac{N_p \times I_p}{I_s}$$

$$N_s = \frac{1 \times 300}{0.5}$$

$$= 600 \text{ turns}$$

Transformer losses

Ideally, the power input to a transformer ($I_p \times V_p$) should equal the power output ($I_s \times V_s$). However, there are power losses which reduce the efficiency. These losses are copper, eddy current and hysteresis losses.

Copper (I^2R) loss

Current flowing in the copper windings causes a heating loss.

Eddy current loss

This loss is caused by alternating currents which are induced magnetically in the core. They are reduced by laminating (*see* Chapter 5).

Hysteresis loss

This is an energy loss due to the changing magnetism in the core.

If we take a sample of unmagnetized iron *(Diagram 71)*, wind a coil on it and pass a current through the winding, the core will become magnetized. The density of the flux will depend on the current and the number of turns. The produce of the current and the turns is called the *magnetizing force (H)*.

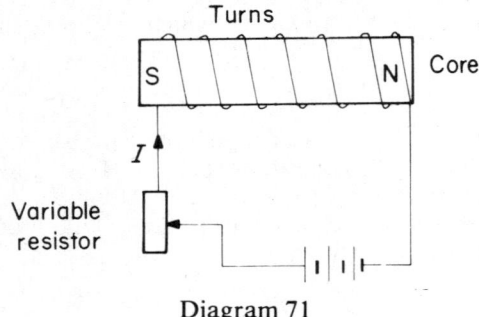

Diagram 71

Clearly, once the coil has been wound it is only the current that is variable, and if it is increased the core becomes magnetized. A graph of this effect is shown in *Diagrams 72 a-e*. The current in the coil is increased from zero to saturation point. Beyond this, an increase in current does not increase the magnetism *(Diagram 72a)*. If the current is now decreased to zero *(see Diagram 72b)*, its path is along (a,c) not (a,o), leaving the core slightly magnetized. This remaining magnetism (o,c) is called the residual magnetism or *remanence*.

If the polarity of the supply is now reversed, and the current increased again the current follows (c,d) *(see Diagram 72c)*. Clearly some force has been used to reduce the remanence to zero (o,d). This force is called the *coercive force*. This would not have occurred if the current had followed the original route (o,a). Hence, energy has been used to overcome the remanence. This is an *energy loss*.

If the current is further increased in the same direction, saturation point will be reached again (e) and the core will have reversed polarity.

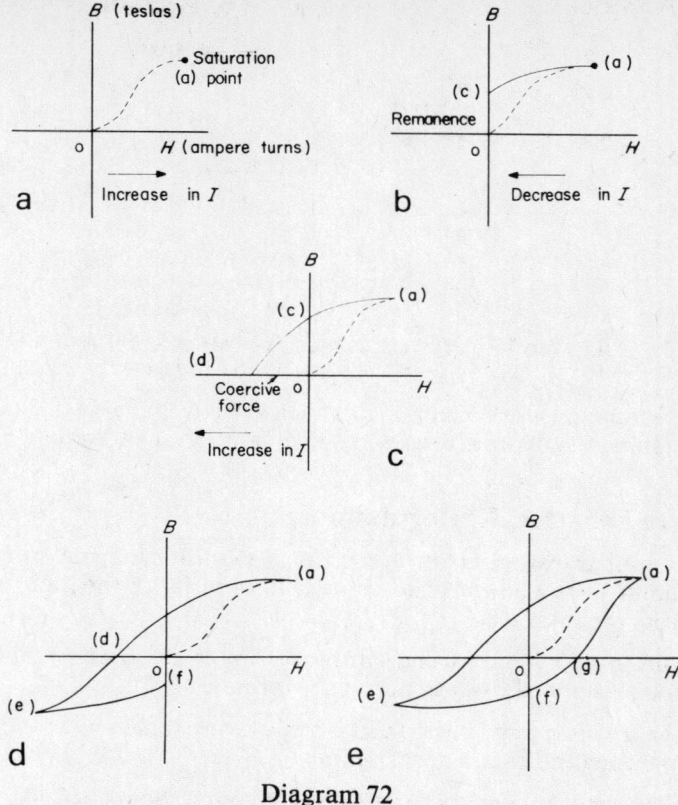

Diagram 72

Decreasing the current to zero will result in a remanence (o,f) which has to be overcome *(Diagram 72d)*.

Another reversal of polarity and an increase in current will result in a coercive force (o,g) being used (energy loss). A further increase in current will bring the curve back to (a) *(Diagram 72e)*.

This complete curve is called the *hysteresis loop*.

If the supply to the coil were alternating current, the polarity would be changing constantly and there would be a continual energy loss. It can be minimized to some extent by using a silicon steel core, the remanence of which is easily overcome *(Diagram 73)*.

Rectifiers

Rectifiers are devices which change a.c. into d.c. The modern rectifier is a comparatively small device (similar to the transistor), and is mainly

62 Cells, batteries and transformers

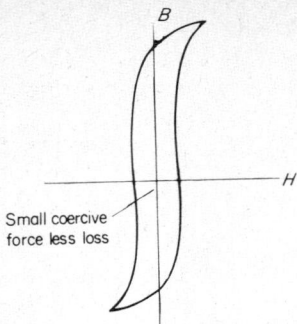

Diagram 73   Hysteresis loop for transformer core

used in installation work in appliances such as battery chargers and/or alarm circuits. Rectifiers are discussed in greater detail in Volume 3.

Points to note (I.E.E. Regulations)

1. Step-up transformers forming part of a consumer installation should have a linked switch provided for isolating the transformer from the supply.
2. Portable appliances using extra-low voltage, i.e. train sets should not be supplied from an auto-transformer.
3. The common terminal of an auto-transformer should be connected to the neutral conductor *(Diagram 69)*.
4. Hand-held appliances likely to be used out of doors or in damp situations should be supplied from a double-wound transformer with a reduced secondary voltage. It is also good practice to earth the centre point of the secondary winding *(Diagram 74)*.
   The recommended secondary voltage is 110 V. Any earth fault on the appliance would result in the user experiencing only 55 V.
5. Supplies for shavers in bathrooms must be obtained from a special shaver unit. This unit has a transformer, the secondary winding of which is completely isolated from the primary and from earth.

Bell, call, fire and burglar alarm systems

This type of system is usually supplied by an extra-low voltage (up to 50 V) although the operating voltage may be supplied via a transformer, whose primary may be at a low voltage.

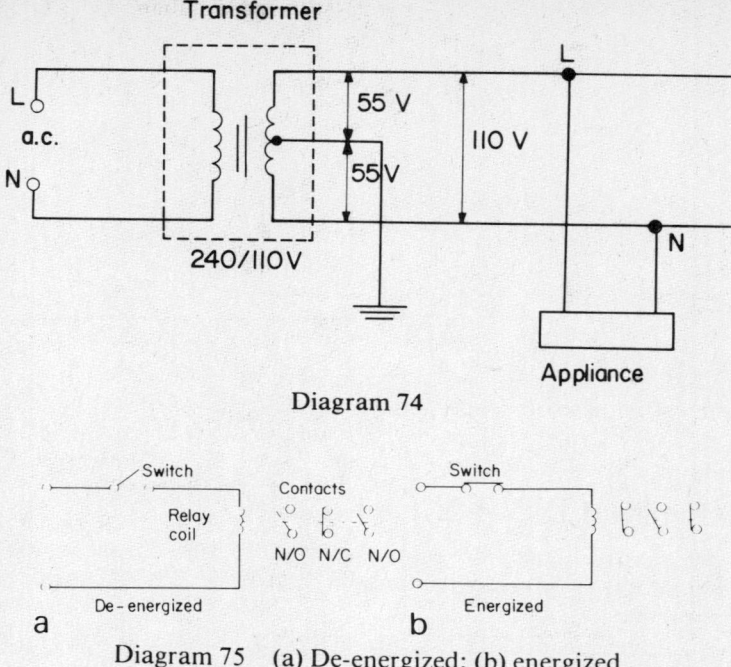

Diagram 74

Diagram 75 (a) De-energized; (b) energized

Before each of these systems is discussed, it is perhaps best to establish a convention with regard to relays. A *relay* is an electromagnet which causes pairs of contacts to make or break, when it is energized. All diagrams should show the relay *de-energized;* the contacts are then said to be in their *normal* position *(Diagram 75).*

Another important point to note is the difference between a circuit or schematic diagram and a wiring diagram. A circuit diagram shows how the system *functions,* and relay contacts, switches, and accessories are shown on a diagram in a position most convenient for drawing and understanding. A wiring diagram shows how the system is to be *wired,* and all components of the circuit should be shown in their correct places *(Diagram 76a* and *76b).*

The 'hold-on' circuit shown in *Diagram 76a* is most important especially in fire and burglar alarm systems. It operates as follows: With switch S closed R is still not energized. By pushing the button B coil R is energized and the normally open contact C will close giving another route for the supply to reach the relay coil. When the button is released the relay will 'hold-on' through its own contact. If switch S were opened, the relay would de-energize and contact C would open, but

64 Cells, batteries and transformers

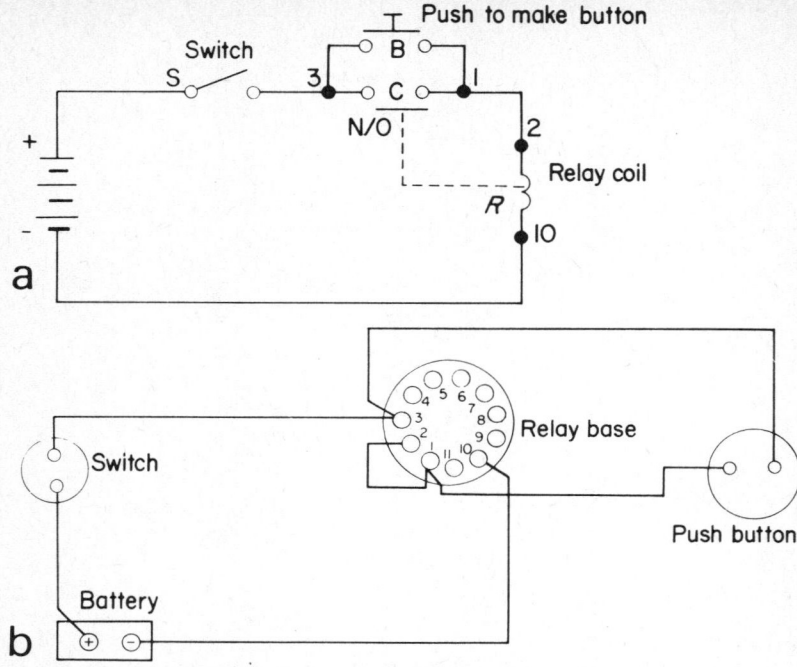

Diagram 76 (a) Simple 'hold-on' circuit; (b) wiring diagram

closing S again would not re-energize the relay. Button B would need to be pushed again to re-energize the relay.

Bell and call systems

These systems and variations of them are widely used in hotels and hospitals. *Diagram 77* shows a basic system.

When a remote push button is operated, its associated relay is energized and an indicator flag drops into a window in the indicator board. As the bell is in series with all of the pushes it will ring when any button is pushed.

The indicator flag is usually latched in position mechanically, and drops into view when the relay is energized. Resetting the flag can be achieved by a mechanical or electrical reset.

Fire and burglar alarm systems

Most large, modern alarm systems are extremely sophisticated, using the techniques of micro-electronics, and as such, are outside the scope of the

Cells, batteries and transformers 65

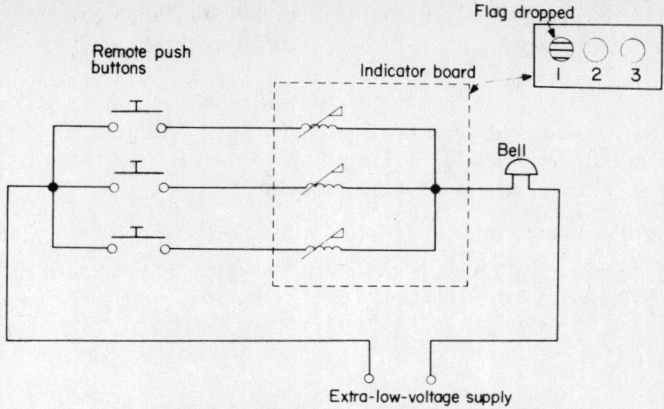

Diagram 77   Basic call system

installation electrician. However, the basic principles of operation are the same, regardless of the size or complexity of the system. We will discuss the simplest forms using relays.

Open-circuit alarm system

The object of this system is to cause a relay to be energized, which, in turn, will energize an alarm bell when a call point is operated. A call point can be a micro-switch, etc. on a door or window (burglar alarm) or a heat detector or 'break glass' point, etc. (fire alarm) *(Diagram 78)*.

When the call point is operated the relay is energized, closing both of the N/O contacts. One set of contacts operates the alarm circuit, the other enables the relay to be held on. Returning the call point to its original position will not permit the relay to de-energize (especially

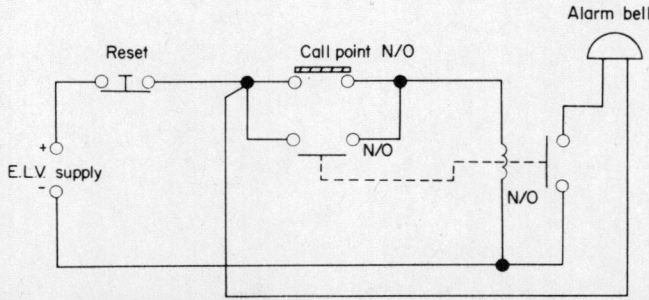

Diagram 78   Open-circuit alarm

important in burglar alarm systems). The system can be reset (relay de-energized) by interrupting the supply to the coil by pressing the reset button.

Extra call points are wired in parallel with the first.

The main disadvantage of this system is that a fault in the relay or a broken or cut wire, rendering the system inoperative, will not be evident until after a fire or a burglary has taken place.

Closed-circuit alarm system

The principle of this system is to de-energize a relay by the operation of a call point *(Diagram 79).*

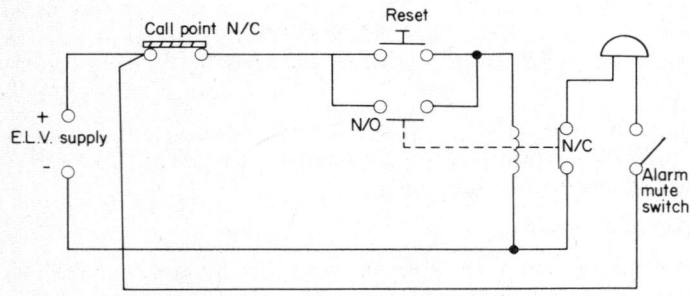

Diagram 79   Closed-circuit alarm

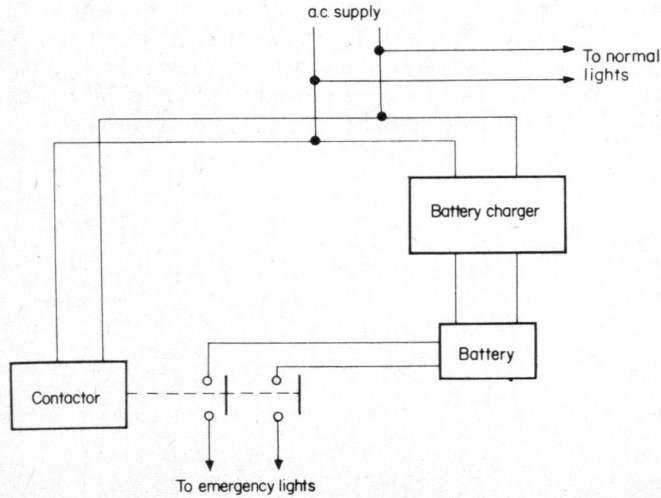

Diagram 80   Non-maintained system

Cells, batteries and transformers   67

In order to prepare the system for operation, the reset button is pushed and the relay is held on through its own N/O contact (provided that the call point is closed). The alarm circuit may now be switched on. An interruption of the relay supply, caused by opening the call point, will de-energize the relay and the N/C contacts will complete the alarm bell circuit. The systems can only be returned to an energized state by the closure of the call point *and* the reset button.

Extra call points are wired in series with the first.

The great advantage of this system is that is is self-monitoring, in that any break in the wiring, failure of the relay or malfunctioning of a call point will cause the alarm to sound.

Fire and burglar alarm systems, especially the closed-circuit type, are best supplied from a central battery with built-in charging facilities. A mains operated system (via a step-down transformer and rectifier) has a problem because a mains failure causes the alarms to sound when the supply is restored. If this happens during the night much inconvenience is caused.

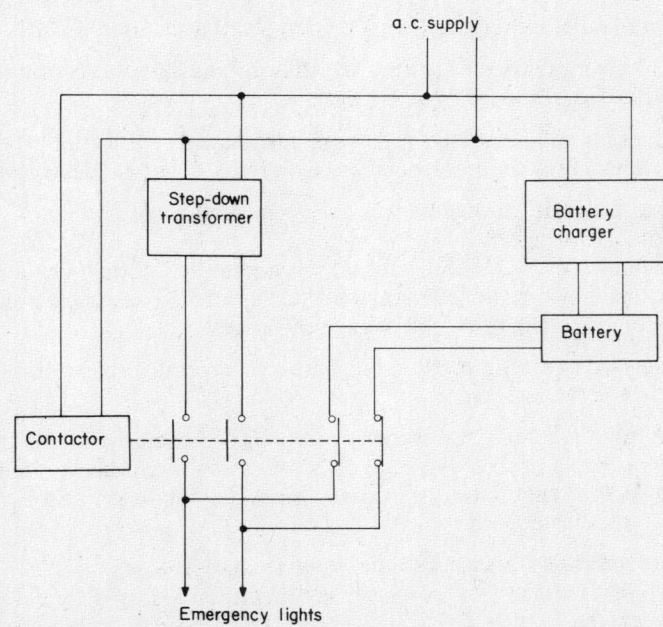

Diagram 81   Maintained system

Emergency lighting

One important use for the secondary cell is to provide alternative supplies to a lighting installation in the event of a mains failure.

There are two main systems of emergency lighting: the non-maintained system, and the maintained system.

In the system shown in *Diagram 80* the emergency lights are only energized when a mains failure occurs. The contactor will de-energize, the contacts close and the lights will be fed from the battery.

In a maintained system *(Diagram 81)* the lights are fed from two sources, a.c. and d.c. When the mains fail, the contactor de-energizes and the contacts change the supply to the lights over to the battery.

Both schemes, in modern practice, use electronics to make the change-over from mains to battery but the contactor system is suitable for showing the principle of operation.

Questions on Chapter 4

1. Explain the difference between a primary and a secondary cell.
2. Describe two methods of testing the state of charge of a secondary cell. What figures would you expect?
3. How does an alkaline cell differ from a lead–acid cell? What effect does discharging an alkaline cell have on the S.G. of its electrolyte?
4. (a) What is the difference between the e.m.f. of a cell and its terminal voltage?
   (b) A cell of e.m.f. 2 V and internal resistance $0.15\ \Omega$ delivers a current of 3 A to an external load. Calculate the resistance of the load and the terminal voltage of the cell.
5. State the two common methods of battery charging and explain the difference between them.
6. (a) Explain the action of a transformer. (b) A double-wound transformer has a primary voltage of 240 V and a secondary voltage of 110 V. If there are 720 primary turns, calculate the number of turns on the secondary.
7. With the aid of a diagram explain what an *auto-transformer* is. What are the restrictions on its use as recommended by the I.E.E. Regulations?
8. List the losses which occur in a transformer. What steps can be taken to overcome them?

9. Explain with the aid of a diagram a closed-circuit alarm system. What advantage does this system have over an open-circuit type?
10. What is meant by (a) a non-maintained and (b) a maintained emergency lighting system?

CHAPTER 5

Electrical Industries

Designing an electrical installation

Every electrical installation, no matter how small, has to be designed, whether mentally or by preparing a specification and/or a plan/drawing.

Even for the installation of a single socket outlet, a decision must be made as to the best position, the most economic cable route and the size of cable, and therefore a design has to be considered.

In order to illustrate the procedure that has to be adopted in the design of an electrical installation, let us consider an installation in a new domestic dwelling.

Specifications and drawings

Usually a building firm invites tenders from installation contractors for the wiring of a property, and the specifications and drawings are sent to all the interested parties.

The specification indicates the quantity of sockets, lighting points and other fixed accessories required for each room, and sometimes the preferred manufacturer. It may also indicate the type and size of cable to be used (*see* specimen specification p.72-76).

The architect's drawings normally show elevations and plan views, with details of recommended positions for all accessories. (The plan between p.76 and 77 omits these latter details.)

It is from these two sources of information, specification and drawing, that a design for the electrical installation can be prepared and a competitive tender submitted.

The design team

This may consist of one or more people, depending on the size of the contracting firm and the complexity of the installation. The example in question (a domestic dwelling) would require only one designer. He will

Phone: Waterlooville 2583

REGISTERED HOUSE-BUILDER

Your Ref:
Our Ref: FB/HI

F. A. Smith (Waterlooville) Ltd.

Building Contractor
Registered Office
88 Jubilee Road, Waterlooville, Hants. PO7 7RE
Company No 662680 Registered in England

3 May 1979

M Jones & Co
34 Queensway
Anytown
England

Dear Sirs

Re: <u>Electrical Installation at Hunters Meadow Estate</u>

We should be pleased to receive by return your lowest price for materials/subcontract work as detailed below, delivered to/executed at the above project.

Workmanship and materials must comply with the appropriate clauses of the current edition of the registered House-Builder's Handbook Part II: Technical.

As the main developers for this project we have the order to place for these items.

Yours faithfully
F A Smith (Waterlooville) Limited

F Bloggs

Fred Bloggs
<u>Buyer</u>

E/O for connecting central heating.

<u>Wiring</u> To be carried out in PVC cable; lighting wired in 1.0 MM: ring main in 2.5 MM

<u>Sheathing</u> PVC sheathing to be fixed on walls under plaster.

<u>Boxes</u> Metal boxes to be used for all fittings on plastered walls.

<u>Fittings</u> To be of cream flush Crabtree or equivalent; 13 amp socket outlets for ring main.

HUNTERS MEADOW ESTATE, ACACIA AVENUE, ANYTOWN

TYPE "B" 3 BED LINKED

This spacious Link/Detached three-bedroomed house with attractive elevations has a larger than average lounge with separate dining area, good-sized bedrooms and downstairs cloakroom making the ideal family home.

Heating is by means of gas-fired central heating and an electric fire in hardwood surround to lounge. Thermoplastic floor tiles to choice from our standard range for the whole of the ground floor.

ACCOMMODATION

Lounge	19' 0" x 13' 0"
Dining room	9' 0" x 10' 9"
Kitchen	9' 9" x 10' 9"
Hall	5' 0" x 5' 0"
Cloaks	5' 0" x 3' 9"
Bedroom 1	12' 0" x 12' 0"
Bedroom 2	9' 0" x 11' 9"
Bedroom 3	9' 0" x 7' 0"
Bathroom	6' 6" x 6' 9"
Airing cupboard	3' 0" x 2' 0"

EXTERNAL DOORS Are softwood glazed in two squares to match windows

GARAGE DOORS Metal up and over

INTERNAL DOORS Sapele hardwood flush

CEILINGS Artex stipple finish

DOOR AND WINDOW FURNITURE In satin anodized aluminium

CURTAIN BATTENS to all windows

LOUNGE

2 Ceiling light points
3 - 13 amp socket outlets
1 T.V. aerial socket outlet with down lead from roof space
1 Point for electric fire
2 Radiators

TYPE "B" 3 BED LINKED (Continued)

DINING ROOM

1 Ceiling light point
Twin socket outlet
1 Radiator

KITCHEN

1 Fluorescent ceiling light
3 - 13 amp socket outlets
1 Cooker control with 1 additional socket incorporated
1 Immersion heater switch
1 - 13 amp point for gas boiler
Wall-fitted gas boiler for central heating
3 Gas points
Double drainer stainless steel sink unit mixer taps and cupboard under
1 Large base unit
1 Broom cupboard
1 High level cupboard

HALL

1 Ceiling light point

CLOAKROOM

1 Ceiling light point
1 Low-level W.C. suite
1 Corner hand basin

GARAGE

Metal up and over door
1 Ceiling light point
1 - 13 amp socket outlet
Electric and gas meters

LANDING

1 Ceiling light point
1 - 13 amp socket outlet
1 Full height airing cupboard with slatted shelves

TYPE "B" 3 BED LINKED (Continued)

BEDROOM 1

1 Ceiling light point
2 - 13 amp socket outlets
1 Radiator

BEDROOM 2

1 Ceiling light point
2 - 13 amp socket outlets
1 Radiator

BEDROOM 3

1 Ceiling light point
2 - 13 amp socket outlets
1 Radiator

BATHROOM

1 Ceiling light point
Coloured bathroom suite comprising low-level suite, pedestal basin and 5' 6" bath
1 Chromium-plated towel rail (electric)
Half-tiled walls to match suite colour
Bathroom cabinet with mirror front
1 Shaver point

EXTERNAL

Tarmac drive to garage
Paved concrete paths to front and back entrance 2' 6" wide
Turfed open-plan front garden
Rear and side boundaries are defined by various means to suit the development as a whole, e.g. screen walls, chain link fences, etc.
Interwoven fencing panels erected adjacent to the property

EXTERNAL COLOUR SCHEME

Woodwork is generally white but details are featured in colours which are designed for the development as a whole.

TYPE "B" 3 BED LINKED (Continued)

INTERNAL COLOUR SCHEME

A limited choice of internal decoration is available from our standard range.

VARIATIONS

The house has been designed as an integrated unit and as a general rule it is not possible to accept variations.

THIS INFORMATION IS GIVEN AS A GUIDE TO PROSPECTIVE PURCHASERS. THE DETAILS MAY VARY FROM ONE PLOT TO ANOTHER AND FULL INFORMATION CAN BE OBTAINED FROM OUR SALES PERSONNEL. THIS SPECIFICATION DOES NOT FORM PART OF ANY CONTRACT AND MAY BE REVISED WITHOUT PRIOR NOTICE.

need to be experienced, not only in electrical work, but in other trades and be able to interpret the architect's drawings. He will also, in the absence of any positive information, provisionally locate all electrical accessories using the symbols given in B.S. 3939 *(Table 1)*. These symbols are extremely important, especially to the installation team.

The installation team

Once again, the team will comprise one or more people, depending on the contract. For the work in this example, two people would normally be employed; an approved electrician and perhaps an apprentice.

A good electrician will have a reasonable knowledge of the basic principles of the other trades involved in the building and servicing of the house, and in this respect, he will often need to discuss, competently, problems which may occur due to the location of his materials in relation to those of different trades. He must, of course, be familiar with the symbols in B.S. 3939, and be able to communicate intelligently with the client or his representative.

The client

The client, in this case usually the builder, has to rely on the expertise of the design and installation teams; it is therefore important that a good liaison be maintained throughout the duration of the contract.

It is the satisfied client who places further contracts with the installation firm.

Working relationships

Working with the building trade on site can present its own problems not found in say, private re-wiring. Good working relationships with other trades go a long way to overcoming any problems. The phrase 'good working relationships' does not mean just having a pleasant attitude to other workers; it involves liaising with them either directly or via a site foreman, and endeavouring not to hinder their work progress. A schedule of intended work with proposed dates can help a great deal. There is nothing more damaging to working relationships than, for example, a plastering team arriving on site to find that the cable drops are not complete and the electrician is not on site.

Safety and welfare

Regardless of the location, any working situation requiring the use of tools, is potentially hazardous. Care must be taken in the selection of the correct tool for the job, and its use.

Table 1 Symbols for architectural and installation diagrams (taken from B.S. 3939: Section 27)

No.	Description	Symbol
27.1	**Routing of conductors**	
27.1.1	Wiring: general symbol	———
27.1.2	Wiring on the surface	⊓ ⊓
27.1.3	Wiring under the surface	⊓ ⊓
27.1.4	Wiring in conduit	—o—
27.1.5	Wiring in duct or trunking	—□—
27.1.6	Wiring going upwards	∕∘
27.1.7	Wiring going downwards	∕•
27.1.8	Wiring passing vertically through a room	∕• ∕∘
27.1.9	Joint or junction box Example with three outlets *Note:* For a joint box the dot at the connection joint must be shown	⊕
27.2	**Lighting points or lamps**	
27.2.1	Lighting point or lamp: general symbol	✕

Table 1 (continued)

No.	Description	Symbol
	Note: The number, power and type of the light source should be specified *Example:* Three 40 watt lamps	✕ 3 x 40 W
27.2.2	Lamp or lighting point: wall mounted	
27.2.3	Emergency (safety) lighting point	
27.2.4	Lighting point with built-in switch	
27.2.5	Lamp fed from variable voltage supply	
27.2.6	Projector or lamp with reflector	
27.2.7	Spotlight	
27.2.8	Single fluorescent lamp	
27.2.9	Group of three fluorescent lamps *Example:* Simplified representation	3 x 40 W
27.2.10	Auxiliary apparatus for discharge lamp *Note:* Only used when the	

Table 1 (continued)

No.	Description	Symbol
	auxiliary apparatus is separated from the lamp fixture	
27.2.11	Illuminated sign (annotation as required) *Note:* The annotation may include an arrow to indicate direction	
27.2.12	Illuminated emergency or safety sign	
27.2.13	Signal lamp	⊗
27.3	**Switches and switch outlets**	
27.3.1	Single-pole, one-way switch *Note:* Number of switches at one point may be indicated	
27.3.2	Two-pole, one way switch	
27.3.3	Three-pole, one way switch	
27.3.4	Cord-operated single-pole one way switch	
27.3.5	Two-way switch	

Table 1 (continued)

No.	Description	Symbol
27.3.6	Intermediate switch	
27.3.7	Time switch	
27.3.8	Switch with pilot lamp	
27.3.9	Period-limiting switch	
27.3.10	Regulating switch e.g. Dimmer	
27.3.11	Push button	
27.3.12	Luminous push button	
27.4	**Socket outlets**	
27.4.1	Socket outlet (mains): general symbol *Note:* In UK practice this general symbol normally implies the presence of an earthing contact. Exceptions to this rule should be indicated by a note, e.g. shaver outlet	
27.4.2	Switched socket outlet	

Table 1 (continued)

No.	Description	Symbol
27.4.3	Socket outlet with interlocking switch	
27.4.4	Socket outlet with pilot lamp	
27.4.5	Multiple socket outlet *Example:* for 3 plugs	
27.5	**Control and distribution**	
27.5.1	Main control or intake point	
27.5.2	Distribution board or point *Note:* The circuits controlled by the distribution board may be shown by the addition of an appropriate qualifying symbol or reference	
	Examples Heating (See also 27.6.3)	
	Lighting (See also 27.2.1)	
	Ventilating (See also 27.6.2)	
27.5.3	Main or sub-main switch	

Table 1 (continued)

No.	Description	Symbol
27.5.4	Contactor	
27.5.5	Integrating meter	
27.5.6	Starter	
27.5.7	Changeover switch	
27.5.8	Transformer	
27.5.9	Consumer's earthing terminal	●E
27.6	**Fixed apparatus and equipment including fire alarm and security devices**	
27.6.1	Electrical appliance: general symbol *Note:* If necessary use designations to specify type	
27.6.2	Fan	

Table 1 (continued)

No.	Description	Symbol
27.6.3	Heater Type to be specified	⊣▭▭▭▭
27.6.4	Motor: general symbol	(M)
27.6.5	Generator: general symbol	(G)
27.6.6	Thermostat: block symbol	[t˙]
27.6.7	Restricted-access push button	[○]
27.6.8	Restricted-access push button for fire alarm	[[○]]
27.6.9	Automatic fire detector	[!]
27.6.10	Watchman system device or key-operated switch	[♦]
27.6.11	Bell	⌒
27.6.12	Indicator panel N = number of ways	[⊖] N

Table 1 (continued)

No.	Description	Symbol
27.6.13	Clock	
27.7	**Telecommunication apparatus including radio and television**	
27.7.1	Telephone call point *Note:* Special services may ber indicated by appropriate references	
27.7.2	Manual switchboard: general symbol	
27.7.3	Automatic telephone exchange equipment	
27.7.4	Socket outlet for telecommunication: general symbol	
	Examples: Television	TV
	Radio	R
	Sound	S
27.7.5	Teleprinter	
27.7.6	Aerial	

Table 1 (continued)

No.	Description	Symbol
27.7.7	Earth	⏚
27.7.8	Radio or television receiver. Service to be specified as in 27.7.4	□
27.7.9	Amplifier	▷
27.7.10	Microphone	◐
27.7.11	Loudspeaker	◁

Symbols from B.S. 3939: Section 27 are reproduced by permission of the British Standards Institution, 2 Park Street, London W1A 2BS, from whom complete copies of the Standard may be obtained.

Building sites are particularly dangerous areas and special care must be taken when working on such sites. Safety helmets and protective shoes should be worn, and where brickwork has to be channelled out, the use of goggles is advisable.

Portable electric tools present a major hazard area on sites, and to overcome the danger of serious electric shock from faulty portable equipment, the I.E.E. Regulations recommend, that a reduced voltage system be used, the supply should be obtained from a double-wound transformer having a reduced secondary voltage. The usual arrangement is a double-wound transformer (240 V/110 V) with a secondary winding centre tapped to earth. This ensures that any fault, P—N or N—E on the portable appliance, will result in the operator only receiving 55 V to earth.

The Regulations outline recommendations for installations on construction sites, and for temporary installations.

A knowledge of simple first aid is a great asset. But in the absence of such skills, it is essential to know which, if any, of the personnel on sites are qualified first aiders.

Existing installations

The installation of extra circuits, maintenance or repair of faulty equipment in an existing installation requires special care, as, in many cases, live supplies are present. When working on such installations the following safety measures should be taken:

1 Switch off the supply to the circuit.

2 Remove fuses and keep on person or locked away.

3 Check, with an approved tester, that the apparatus is dead. This is especially important when the isolation has been carried out by others.

Customer relationships

In a private dwelling requiring a re-wire or simply additional lighting or power points, the electrician(s) must have the correct attitude to the customer and his property; after all, it is the customer who is paying for the work and he, justifiably, expects the best possible service. The following are some examples of conduct when working in other people's property:

DO Be polite under *all* circumstances.

DO Be presentable in dress and manner.

DO   Be tidy, clear away all unnecessary debris, replace furniture to its original position.

DO   Consult the customer if positions of accessories are not clearly defined.

DON'T   Use bad language.

DON'T   Place tools on furniture.

DON'T   Use furniture as steps.

DON'T   Leave without informing customer (unless he is out).

DON'T   Leave floorboards and/or carpets unsecured when work is completed.

Definitions

The following definitions indicate the sense in which the terms defined are used in these Regulations. Some of these definitions are aligned with those given in B.S. 4727 – 'Glossary of electro-technical, power, telecommunication, electronics, lighting and colour terms'. Other terms not defined herein are used in the sense defined in B.S. 4727.

Accessory

A device, other than current-using equipment, associated with such equipment or with the wiring of an installation.

Ambient temperature

The temperature of the air or other medium where the equipment is to be used.

Appliance

An item of current-using equipment other than a luminaire or an independent motor.

Arm's reach

A zone of accessibility to touch, extending from any point on a surface where persons usually stand or move about, to the limits which a person can reach with his hand in any direction without assistance.

Barrier

A part providing a defined degree of protection against contact with live parts, from any usual direction of access.

Basic insulation

Insulation applied to live parts to provide basic protection against electric shock.

Note: Basic insulation does not necessarily include insulation used exclusively for functional purposes.

Bonding conductor

A protective conductor providing equipotential bonding.

Bunched

Cables are said to be bunched when two or more are contained within a single conduit, duct, ducting, or trunking or if not enclosed, are not separated from each other.

Cable coupler

A means enabling the connection, at will, of two flexible cables. It consists of a connector and a plug.

Cable ducting

A manufactured enclosure of metal or insulating material, other than conduit or cable trunking, intended for the protection of cables which are drawn-in after erection of the ducting, but which is not specifically intended to form part of a building structure.

Caravan

Any structure designed or adapted for human habitation which is capable of being moved from one place to another (whether by being towed or being transported on a motor vehicle or trailer) and any other motor vehicle so designed or adapted but not including—
 (a) any railway rolling stock which is for the time being on rails forming part of a railway system, or
 (b) any tent.

Note: For multi-unit structures see the amendment to the definition for caravans in the Caravan Sites Act 1968.

Caravan site

Land on which a caravan is stationed for the purposes of human habitation and land which is used in conjunction with land on which a caravan is so stationed.

Cartridge fuse link

A device comprising a fuse element or several fuse elements connected in parallel enclosed in a cartridge usually filled with an arc-extinguishing medium and connected to terminations. The fuse link is the part of a fuse which requires replacing after the fuse has operated.

Circuit

An assembly of electrical equipment supplied from the same origin and protected against overcurrent by the same protective device(s). For the purposes of Chapter 52 of these Regulations, certain types of circuit are categorised as follows:

Category 1 circuit – A circuit (other than a fire alarm or emergency lighting circuit) operating at low voltage and supplied directly from a mains supply system.

Category 2 circuit – With the exception of fire alarm and emergency lighting circuits, any circuit for telecommunication (e.g. radio, telephone, sound distribution, intruder alarm, bell and call, and data transmission circuits) which is supplied from a safety source complying with Regulation 411-3.

Category 3 circuit – A fire alarm circuit or an emergency lighting circuit.

Circuit breaker

A mechanical switching device capable of making, carrying and breaking currents under normal circuit conditions and also of making, carrying for a specified time, and breaking currents under specified abnormal circuit conditions such as those of short circuit.

Note: A circuit breaker is usually intended to operate infrequently, although some types are suitable for frequent operation.

Circuit protective conductor

A protective conductor connecting exposed conductive parts of equipment to the main earthing terminal.

Class I equipment

Equipment in which protection against electric shock does not rely on basic insulation only, but which includes means for the connection of exposed conductive parts to a protective conductor in the fixed wiring of the installation.

Note: For information on classification of equipment with regard to means provided for protection against electric shock see B.S. 2754.

Class II equipment

Equipment in which protection against electric shock does not rely on basic insulation only, but in which additional safety precautions such as supplementary insulation are provided, there being no provision for the connection of exposed metalwork of the equipment to a protective conductor, and no reliance upon precautions to be taken in the fixed wiring of the installation.

Note: For information on classification of equipment with regard to means provided for protection against electric shock see B.S. 2754.

Confined conductive location

A location having surfaces which are mainly composed of extraneous conductive parts and which are of such dimensions that movement is restricted to such an extent that contact with surfaces is difficult to avoid (e.g. in a boiler).

Connector

The part of a cable coupler or of an appliance coupler which is provided with female contacts and is intended to be attached to the flexible cable connected to the supply.

Current-carrying capacity of a conductor

The maximum current which can be carried by a conductor under specified conditions without its steady state temperature exceeding a specified value.

Current-using equipment

Equipment which converts electrical energy into another form of energy, such as light, heat, or motive power.

Danger

Danger to health or danger to life or limb from shock, burn, or injury from mechanical movement to persons (and livestock where present), or from fire, attendant upon the use of electrical energy.

Design current (of a circuit)

The magnitude of the current intended to be carried by the circuit in normal service.

Direct contact

Contact of persons or livestock with live parts which may result in electric shock.

Double insulation

Insulation comprising both basic insulation and supplementary insulation.

Duct

A closed passage way formed underground or in a structure and intended to receive one or more cables which may be drawn in.

Ducting

See Cable ducting.

Earth

The conductive mass of the Earth, whose electric potential at any point is conventionally taken as zero.

Earth electrode

A conductor or group of conductors in intimate contact with, and providing an electrical connection to, Earth.

Earth electrode resistance

The resistance of an earth electrode to Earth.

Earth fault loop impedance

The impedance of the earth fault current loop (phase to earth loop) starting and ending at the point of earth fault.
   Note: See Appendix 15 for a description of the constituent parts of an earth fault current loop.

Earth leakage current

A current which flows to Earth, or to extraneous conductive parts, in a circuit which is electrically sound.
   Note: This current may have a capacitive component including that resulting from the deliberate use of capacitors.

Earthed concentric wiring

A wiring system in which one or more insulated conductors are completely surrounded throughout their length by a conductor, for example a sheath, which acts as a PEN conductor.

Earthing conductor

A protective conductor connecting a main earthing terminal of an installation to an earth electrode or to other means of earthing.

Electric shock

A dangerous pathophysiological effect resulting from the passing of an electric current through a human body or an animal.

Electrical equipment (abbr. *Equipment*)

Any item for such purposes as generation, conversion, transmission, distribution or utilisation of electrical energy, such as machines, transformers, apparatus, measuring instruments, protective devices, wiring materials, accessories, and appliances.

Electrical installation (abbr: *Installation*)

An assembly of associated electrical equipment to fulfil a specific purpose and having certain co-ordinated characteristics.

Electrically independent earth electrodes

Earth electrodes located at such a distance from one another that the maximum current likely to flow through one of them does not significantly affect the potential of the other(s).

Electrode boiler (or electrode water heater)

Equipment for the electrical heating of water or electrolyte by the passage of an electric current between electrodes immersed in the water or electrolyte.

Emergency switching

Rapid cutting-off of electrical energy to remove any hazard to persons, livestock, or property which may occur unexpectedly.

Enclosure

A part providing an appropriate degree of protection of equipment against certain external influences and a defined degree of protection against contact with live parts from any direction.

Equipment

(abbr. *see Electrical equipment*)

Equipotential bonding

Electrical connection putting various exposed conductive parts and extraneous conductive parts at a substantially equal potential.

Exposed conductive part

A conductive part of equipment which can be touched and which is not a live part but which may become live under fault conditions.

External influence

Any influence external to an electrical installation which affects the design and safe operation of that installation.

Extraneous conductive part

A conductive part liable to introduce a potential including earth potential and not forming part of the electrical installation.

Factory-built assembly (of LV switchgear and control gear)

An assembly built and assembled under the responsibility of the manufacturer, and conforming to an established type of system, without deviations likely to significantly influence the performance, from the typical assembly verified to be in accordance with the relevant British Standard.

> *Note:* For various reasons, e.g. transport or production, certain steps of assembly may be made in a place outside the factory of the manufacturer of the factory-built assembly. Such assemblies are considered as factory-built provided the assembly is performed in accordance with the manufacturer's instructions, in such a manner that compliance with the relevant British Standard is assured, including submission to applicable routine tests.

Final circuit

A circuit connected directly to current-using equipment, or to a socket outlet or socket outlets or other outlet points for the connection of such equipment.

Fixed equipment

Equipment fastened to a support or otherwise secured in a specific location.

Functional earthing

Connection to earth necessary for proper functioning of electrical equipment.

Fuse element

A part of a fuse designed to melt when the fuse operates.

Fuse link

A part of a fuse, including the fuse element(s), which requires replacement by a new or renewable fuse link after the fuse has operated and before the fuse is put back into service.

Indirect contact

Contact of persons or livestock with exposed conductive parts made live by a fault and which may result in electric shock.

Installation (abbr: *see Electrical installation*).

Instructed person

A person adequately advised or supervised by skilled persons to enable him to avoid dangers which electricity may create.

Insulating floor (or wall)

A floor (or wall) such that, in the event of direct contact with a live part, a person standing on the floor (or touching the wall) cannot be traversed by a shock current flowing to the floor (or wall).

Insulation

Suitable non-conductive material enclosing, surrounding, or supporting a conductor.
   Note: See also the definitions for basic insulation, double insulation, reinforced insulation, and supplementary insulation.

Isolation

Cutting off an electrical installation, a circuit, or an item of equipment from every source of electrical energy.

Live part

A conductor or conductive part intended to be energised in normal use, including a neutral conductor but, by convention, not a PEN conductor.

Luminaire

Equipment which distributes, filters, or transforms the light from one or more lamps, and which includes any parts necessary for supporting, fixing and protecting the lamps, but not the lamps themselves, and, where necessary, circuit auxiliaries together with the means for connecting them to the supply.
   Note: For the purposes of these Regulations a batten lampholder, or a lampholder suspended by a flexible cord, is a luminaire.

Main earthing terminal

The terminal or bar provided for the connection of protective conductors, including equipotential bonding conductors, and conductors for functional earthing if any, to the means of earthing.

Mechanical maintenance

The replacement, refurbishment or cleaning of lamps and non-electrical parts of equipment, plant and machinery.

Neutral conductor

A conductor connected to the neutral point of a system and contributing to the transmission of electrical energy.

Note: The term also means the equivalnt conductor of a d.c. system unless otherwise specified in these Regualtions.

Nominal voltage

See Voltage, nominal.

Obstacle

A part preventing unintentional contact with live parts but not preventing deliberate contact.

Origin of an installation

The position at which electrical energy is delivered to an installation.

Overcurrent.

A current exceeding the rated value. For conductors the rated value is the current-carrying capacity.

Overcurrent detection.

A method of establishing that the value of current in a circuit exceeds a predetermined value for a specified length of time.

Overload current

An overcurrent occurring in a circuit which is electrically sound.

PEN conductor

A conductor combining the functions of both protective conductor and neutral conductor.

Phase conductor

A conductor of an a.c. system for the transmission of electrical energy, other than a neutral conductor.
Note: The term also means the equivalent conductor of a d.c. system unless otherwise specified in these Regulations.

Plug

A device, provided with contact pins, which is intended to be attached to a flexible cable, and which can be engaged with a socket outlet or with a connector.

Point (in wiring)

A termination of the fixed wiring intended for the connection of current-using equipment.

Portable equipment

Equipment which is moved while in operation or which can easily be moved from one place to another while connected to the supply.

Protective conductor

A conductor used for some measures of protection against electric shock and intended for connecting together any of the following parts:

>exposed conductive parts
>extraneous conductive parts,
>the main earthing terminal,
>earth electrode(s),
>the earthed point of the source, or an artificial neutral.

Reinforced insulation

Single insulation applied to live parts, which provides a degree of protection against electric shock equivalent to double insulation under the conditions specified in the relevant standard.

>*Note:* The term 'single insulation' does not imply that the insulation must be one homogeneous piece. It may comprise several layers which cannot be tested singly as supplementary or basic insulation.

Residual current device

A mechanical switching device or association of devices intended to cause the opening of the contacts when the residual current attains a given value under specified conditions.

Residual operating current

Residual current which causes the residual current device to operate under specified conditions.

Resistance area (for an earth electrode only)

The surface area of ground (around an earth electrode) on which a significant voltage gradient may exist.

Ring final circuit

A final circuit arranged in the form of a ring and connected to a single point of supply.

Shock current

A current passing through the body of a person or an animal and having characteristics likely to cause dangerous pathophysiological effects.

Short circuit current

An overcurrent resulting from a fault of negligible impedance between live conductors having a difference in potential under normal operating conditions.

Simultaneously accessible parts

Conductors or conductive parts which can be touched simultaneously by a person or, where applicable, by livestock.
 Notes: 1. Simultaneously accessible parts may be
   live parts,
   exposed conductive parts,
   extraneous conductive parts,
   protective conductors,
   earth electrodes.

   2. This term applies for livestock in locations specifically intended for these animals.

Skilled person

A person with technical knowledge or sufficient experience to enable him to avoid dangers which electricity may create.

Socket outlet

A device, provided with female contacts, which is intended to be installed with the fixed wiring, and intended to receive a plug.
 Note: A luminaire track system complying with B.S. 4533 is not regarded as a socket outlet system.

Space factor

The ratio (expressed as a percentage) of the sum of the overall cross-sectional areas of cables (including insulation and any sheath) to the internal cross-sectional area of the conduit or other cable enclosure in which they are installed. The effective overall cross-sectional area of a non-circular cable is taken as that of a circle of diameter equal to the major axis of the cable.

Spur

A branch cable connected to a ring or radial final circuit.

Stationary equipment

Equipment which is either fixed, or equipment having a mass exceeding 18kg and not provided with a carrying handle.

Supplementary insulation

Independent insulation applied in addition to basic insulation in order to provide protection against electric shock in the event of a failure of basic insulation.

Switch

A mechanical switching device capable of making, carrying and breaking current under normal circuit conditions, which may include specified operating overload conditions, and also of carrying for a specified time currents under specified abnormal circuit conditions such as those of short circuit.
   Note: A switch may also be capable of making, but not breaking, short circuit currents.

Switch, linked

A switch the contacts of which are so arranged as to make or break all poles simultaneously or in a definite sequence.

Switchboard

An assembly of switchgear with or without instruments, but the term does not apply to a group of local switches in a final circuit.
   Note: In the Electricity (Factories Act) Special Regulations 1908 and 1944, the term 'Switchboard' includes a distribution board.

Switchgear

An assembly of main and auxiliary switching apparatus for operation, regulation, protection or other control of electrical installations.

System

An electrical system consisting of a single source of electrical energy and an installation. For certain purposes of these Regulations, types of system are identified as follows, depending upon the relationship of the source, and of exposed conductive parts of the installation, to Earth:

   TN system A system having one or more points of the source of energy directly earthed, the exposed conductive parts of the installation being connected to that point by protective conductors. Three types of TN systems are recognised as follows:

      TN-C system, in which neutral and protective functions are combined in a single conductor throughout the system,

      TN-S system, having separate neutral and protective conductors throughout the system,

      TN-C-S system, in which neutral and protective functions are combined in a single conductor in part of the system.

TT system A system having one point of the source of energy directly earthed, the exposed conductive parts of the installation being connected to earth electrodes electrically independent of the earth electrodes of the source.

IT system A system having no direct connection between live parts and Earth, the exposed conductive parts of the electrical installation being earthed.

Note: See Appendix 3 for further explanation of the significance of the terms used to describe types of system.

Trunking (for cables)

A system of enclosures for the protection of cables, normally of square or rectangular cross section, of which one side is removable or hinged.

Voltage, nominal

Voltage by which an installation (or part of an installation) is designated. The following ranges of nominal voltage (r.m.s. values for a.c.) are defined:

—**Extra low** Normally not exceeding 50V a.c. or 120V d.c., whether between conductors or to Earth.

—**Low** Normally exceeding extra-low voltage but not exceeding 1000V a.c. or 1500V d.c. between conductors, or 600V a.c. or 900V d.c. between conductors and Earth.

Note: The actual voltage of the installation may differ from the nominal value by a quantity within normal tolerances.

Definitions and B.S. 3939 symbols

An appreciation of electrical definitions and B.S. symbols is necessary for any competent person involved in electrical installation work, whether at the design or installation stage.

Requisitions and estimates

Before an estimate can be prepared a requisition or materials list must be compiled. This is where experience is invaluable in choosing the correct, and also the most economic, materials for the job. The dwelling we are considering could be wired in m.i.c.c. cable, but although perfectly acceptable, would involve the client in very high costs. So much so, that the tender for the work would stand little chance of being accepted.

The choice of cable and accessories is in fact quite easy in this case, but in more complex contracts the designer may be faced with extreme conditions, and will be involved in a considerable number of calculations for volt drop and current rating before a choice of materials can be made.

100 Electrical industries

In any event, accurate measurements—either on site or, more usually, from the plan—are essential, in order that the shortest cable routes are found. Once the design or layout of the cables and accessories has been established, material quantities can be assessed.

The following examples illustrate typical calculations for the domestic dwelling.

Diversity

In any installation it is unlikely that every power and lighting point and other appliances will be used simultaneously and so the total current drawn from the supply is unlikely to be the total possible.

The I.E.E. Regulations give the percentages of the total connected load for each circuit. It is from these that an estimated likely maximum load can be calculated, and the size of the main cable established.

Note: **The distribution board must be capable of taking the whole load without the application of diversity.**

For example, from the specification it will be seen that the installation will comprise:

Two — 5 A lighting circuits

Two — 30 A ring final circuits

One — 30 A cooker outlet with 5 A socket

One — 15 A immersion heater circuit

Hence applying diversity:

Lighting

$$\text{assumed load} = 13 \text{ points at } 100 \text{ W each}$$

$$= \frac{13 \times 100}{240} = 5.4 \text{ A}$$

$$\text{Diversity} = 66\% \text{ of } 5.4 = \frac{66 \times 5.4}{100} = 3.56 \text{ A}$$

Cooking:
assuming average size cooker of 12 kW

$$\text{assumed load} = \frac{12\,000}{240} = 50 \text{ A}$$

$$\text{Diversity} = 10 \text{ A} + (30\% \text{ of } 40) + 5 \text{ A}$$

$$= 10 \text{ A} + 12 \text{ A} + 5 \text{ A} = 27 \text{ A}$$

Water heater:
assuming a 3 kW element with thermostat

$$\text{assumed load} = \frac{3000}{240} = 12.5 \text{ A}$$

$$\text{Diversity} = \textbf{None allowed} = 12.5 \text{ A}$$

Ring main:

$$\text{Diversity} = 100\% \text{ of } 30 \text{ A} + 40\% \text{ of } 30 \text{ A}$$
$$= 30 \text{ A} + 12 \text{ A} = 42 \text{ A}$$
$$\text{Total load} = 3.56 + 27 + 12.5 + 42 = 85.06 \text{ A}$$

From the tables in the I.E.E. Regulations it will be seen that the cable size (tails) is 16 mm².

Attention is drawn to 'Maximum demand and diversity', I.E.E. Regulations, Appendix 4.

Voltage drop

Although a cable may be capable of carrying the estimated current of a load, it is important that a check be made on the volt drop along the cable to ensure that it does not exceed 2.5% of the nominal supply voltage.

Example 5.1

From the plan of the house it can be found by measurement that the shortest cable run for the cooker circuit is approximately 9 m. The average size of a domestic cooker is in the region of 12 kW.

$$\text{Full load of cooker} = \frac{1200}{240} = 50 \text{ A}$$

Applying diversity:

$$\text{Estimated load} = 10 + 30\% \text{ of } 40 + 5 = 27 \text{ A}$$
$$\text{Nearest fuse size} = 30 \text{ A}$$

As cable size is related to fuse size, from table 9D2 cable size would be 4 mm².
Check volt drop, i.e. 17 mV/ampere/metre

$$\therefore \text{Volt drop} = \frac{18 \times 27 \times 9}{1000} = 2.67 \text{ V}$$

This is acceptable, as maximum permissible volt drop is 2.5% of 240 = 6 V. Hence the cooker cable would be 4.0 mm^2. In new, unoccupied premises it would be sensible to install 6.0 mm^2 cable, in case a larger cooker were to be installed.

Cost of materials and systems

Having completed all necessary calculations, the type of system and its materials must be considered carefully.

Materials
The type of dwelling we are considering would only require P.V.C. twin and earth cable, with drops to switches etc. run in an oval P.V.C. conduit and buried in the plaster, and run unenclosed under the floor. To run the wiring in single-core P.V.C. cable enclosed in a conduit would be pointless and rather expensive.

Systems
The choice of a wiring system is just as important as the choice of material. The system dictates the material quantity. For example, the specification might call for 10 socket outlets (downstairs). Run as a ring main system, only one 30 A way in the consumer unit. Wiring on a radial system would require a calculation of floor area and an increase in cable size to 4.0 mm².

Materials list
Once a system has been decided on and all the calculations have been completed, a list or requisition of materials can be compiled. The example shown of such a list *(Table 2)* is typical for the dwelling we are concerned with.

To complete the quantity column, information must be obtained from the specification and plan, the cost being obtained from the wholesalers, or manufacturer's catalogue.

Equipment on site
Let us assume that the tender has now been accepted, and work has to commence on site. In the case of a large site, a contracting firm may well have a site hut erected for the storing of materials; for smaller contracts,

equipment would be stored at base and transported as required.

In either event, materials have to be delivered and an accurate check on goods delivered must be made and records must be kept. Starting a job, only to find that items have been incorrectly sent is totally inefficient and does not make for good customer relationships.

The correct procedure is as follows:

1. Keep a copy of the original material order.
2. Delivered goods should be accompanied by an *advice note*. Ensure that there is one.
3. Check that the goods correspond with those stated on the advice note.
4. Check the advice note against the original order.

Security

This presents a major problem on the larger site, where much equipment may be stored. There is not always a nightwatchman, and security patrols cannot maintain a constant vigil.

The only answer is to have good padlocks and an alarm system. Apart from the risk of bulk theft of materials, there is always a danger of smaller amounts being stolen, which have been left at the point of work. The simple remedy for this is: do not leave any tools, materials or other equipment lying around after work. Lock everything away.

Protection of materials

All materials used in installation work should be in perfect condition. This cannot be achieved if the materials are carelessly stored or handled on site. All equipment should be kept away from damp or corrosive conditions and equipment involving delicate mechanisms, i.e. thermostats, contactors, relays, etc. should be stored or handled so as to prevent mechanical damage.

If these recommendations are not observed, it is likely that it will be necessary to return to the installation, after completion, to carry out repairs or replacements.

Disposition of equipment on site

In order that a job runs smoothly and efficiently, the positioning of materials on site must be considered. The correct procedure may be summed up as: *Always ensure that all the relevant tools and materials are taken to the place of work.* In this way needless journeys to and from the site hut or base can be eliminated and the correct amount of time can be spent on the installation.

Table 2 Materials requisition sheet

Description	Manuftr.	Code No.	Price ea.	Qty.	Total cost
Cable					
1.0 mm^2					
1.0 mm^2 3 core					
2.5 mm^2					
6.0 mm^2					
16 mm^2 Tails					
6.0 mm^2 Bonding					
0.85 mm^2 Bell wire					
0.5 mm^2 Lighting flex					
1.5 mm^2 Butyl					
2.5 mm^2					
Low-loss T.V. co-axial					
Oval conduit					
12 mm					
20 mm					
25 mm					
Metal boxes (KO)					
Plaster depth					
25 mm deep 1 gang					
25 mm deep 2 gang					
35 mm deep 2 gang					
Cable clips					
1.0 mm					
2.5 mm					
6.0 mm					

cont'd

Description	Manuftr.	Code No.	Price ea.	Qty.	Total cost
Power					
Switched socket outlet (D)					
Switched socket outlet (S)					
Switched fused spur box					
Cooker control box					
Lighting					
1-gang 1-way plate switch					
1-gang 2-way plate switch					
2-gang 2-way plate switch					
3-gang 2-way plate switch					
Pull cord switch					
Ceiling rose					
Lampholder					
Battenholder					
Other					
Shaver point					
Heated towel rail					
Immersion heater					
Thermostat					
8-way consumer unit					
5 A fuse and base					
15 A fuse and base					
30 A fuse and base					
Earth clamp					
Bell					
Bell transformer					
Bell push					
4 ft. fluorescent					
Fitting					
T.V. outlet					
Surface boxes (S/O)					
Surface boxes (switch)					
Earth sleeving					

Correction factors

The examples on diversity and volt drop previously given were, of course, relatively straightforward. In many cases, however, there are other factors which influence the current-carrying ability of a cable.

As we already know, an increase in conductor temperature causes an increase in conductor resistance and hence a corresponding decrease in the conductor's current-carrying capacity. So, a cable carrying its fully rated current will be warm, and, if it is in an area of high ambient temperature and close to other cables, its temperature will rise further. If it is also protected by a semi-enclosed rewirable fuse, it could take an overload for a considerable length of time. All of these factors have an adverse effect on the cable and must therefore be taken into account when selecting the cable required for a job.

The tables of current rating and voltage drop in the I.E.E. Regulations give correction factors which are applied to the nominal value of the protective device in order to determine the correct cable size for the environmental conditions.

Once the design current of the circuit has been established, a suitable rating for the protection is chosen e.g. a circuit supplying a 4.5 kW heating appliance would have a design current of $4500/240 \times 18.75$ A. Hence the nominal current setting of the protection would be 20 A. The correction factors are then divided into this value and the resulting figure will be the cable current rating.

Example 5.2

A 7 kW shower unit is to be supplied from a single pole switch fuse. The cable (P.V.C./P.V.C. sheathed twin with C.P.C.) is to be buried direct in plaster and clipped to the surface in a loft space avoiding the thermal insulation. The maximum temperature in the roof space will be 40°C and the length of the run will be 15 m. Calculate the minimum cable size if the circuit is protected by (a) a fuse to B.S. 3036 or (b) a circuit breaker to B.S. 3871.

(a) Design current $I_B = \dfrac{P}{V} = \dfrac{7000}{240} = 29.17$ A

$\therefore$ Nominal setting of protection $I_N = 30$ A
Correction factors (see I.E.E. Regulations):

Ambient temperature	0.94
Grouping of cables	N/A
B.S. 3036 fuse	0.725
Thermal insulation	N/A

Hence current carrying capacity of cable, $I_z = \dfrac{I_N}{\text{correction factors}}$

$$= \dfrac{30}{0.94 \times 0.725}$$
$$= 44 \text{ A}$$

∴ cable size (from tables) = 6.0 mm²

and volt drop $= \dfrac{7.3 \times 29.17 \times 15}{1000} = 3.2$ V (acceptable)

(b) as (a) then:
Correction factors:
Ambient temperature only   0.87
Thermal insulation   0.75

∴ $I_z = \dfrac{I_N}{0.87} = \dfrac{30}{0.87} = 34.4$ A

∴ cable size (from tables) = 4.0 mm²

and volt drop $= \dfrac{11 \times 2.17 \times 15}{1000} = 4.8$ V (acceptable)

These calculations establish that in each case a twin cable can be selected for its current carrying capacity and voltage drop. However, such twin cables have a smaller C.P.C. than the phase conductor and a further check must be made that the C.P.C. size is adequate to ensure that there is no risk of shock and that no damage will occur in the event of a fault. This topic is discussed briefly in Volume 3 of this series and in detail in *The IEE Wiring Regulations Explained and Illustrated* by the same author.

There are, of course, other adverse conditions which affect the choice of the type of cable used and indeed the *type* of conduit or trunking. Such conditions may be flammable, explosive, damp or corrosive, and a suitable system comprising materials which are safe to install must be selected.

Flammable and explosive situations

1. A flammable material is one capable of being easily ignited.

2. Typical situations which may be described as flammable and/or explosive would be garages dispensing petrol, chemical works handling and/or storing volatile liquid or vapour or gas, dusty areas such as flour mills, etc.
Cables in such situations should be installed such that there is no risk of mechanical damage which could cause sparking. This is achieved by drawing the cables in solid drawn or seam-welded conduit, or using m.i.c.s. cable or lead-sheathed steel-armoured cable or P.V.C.-insulated armoured cable with P.V.C. sheath.
3. Terminations in flammable and/or explosive situations should be avoided. If this is not possible the terminations should be enclosed in a flameproof fitting.
4. Care must be taken when installing fixed apparatus, i.e. motors, fans, heaters, etc., to ensure that suitable enclosures are used, or other safety measures taken to avoid a hazardous situation arising.

Damp and corrosive situations

1. No metal sheaths or armouring of cable, metal conduit or trunking, or clips or fixing should be installed in a damp or corrosive situation unless they are of a corrosion-resistant material or finish. Contact with other metals with which electrolytic action might occur is to be avoided.
2. Agricultural and horticultural installations are particularly prone to damp and corrosion and special care must be taken to select the correct materials.

Wiring between buildings

The domestic dwelling we have been dealing with in this chapter has a garage attached. However, many premises have remote garages or outhouses which require a supply of electricity. This is not simply a matter of either digging a trench for the cable or hanging it between the buildings. The following points from the Regulations indicate the correct methods.

Points to note (I.E.E. Regulations)

The following types of cable may be used underground:
1. In ducts:
    (a) metal-sheathed and served cable
    (b) m.i.m.s. P.V.C.-sheathed cable

(c) s.w.a. P.V.C.-insulated P.V.C.-sheathed cable

(d) P.V.C.-insulated P.V.C.-sheathed cable or rubber-insulated cable having a heat, oil and flame retardant sheath (h.o.f.r.).

2. In conduits or pipes:

P.V.C.-insulated P.V.C.-sheathed cable, or rubber-insulated cable with h.o.f.r. sheath, provided that the conduit is adequately protected and made of heavy-gauge steel, or galvanized steel pipe, or in a non-metallic conduit.

3 Laid direct in ground:

(a) metal-sheathed armoured and served underground cable

(b) s.w.a. P.V.C.-insulated P.V.C.-sheathed cable

(c) metal-sheathed and served underground cable or m.i.c.s. with P.V.C. sheath, protected by cable covers, except where installed under a permanent surface, and installed at a reasonable depth to avoid damage under normal circumstances.

The following types of cable may be used clipped to an exterior surface and boundary walls, etc *(but not fences)*:

1. Cables run in heavy-gauge, hot dipped, zinc-coated steel conduit or steel pipes having a corrosion-resistant finish.
2. M.i.m.s. cable with black P.V.C. sheath.
3. Armoured cables with black P.V.C. sheath.
4. Lead-sheathed cable.
5. P.V.C.-insulated, black P.V.C.-sheathed or cables run in black P.V.C. high-impact rigid conduit.

The following are some of the types of cable which may be used for overhead wiring:

1. P.V.C.-insulated, P.V.C.-sheathed cable, or rubber insulated cable with h.o.f.r. sheath (black).
2. Cable as listed in (1) above installed in heavy gauge hot dipped zinc-coated steel conduit.
3. Cable supported by a catenary wire.

All of these cables are subject to the maximum span length and height specified in the appropriate tables in the Regulations.

CHAPTER 6

Control and Earthing

The I.E.E. Regulations state that every consumer's installation must have a means of:

1. Isolation
2. Overcurrent protection
3. Earth leakage protection

Isolation

The means of isolation can be achieved by several methods; these include the isolator, switch fuse, fuse switch, consumer unit, and circuit breaker.

Isolator

This is simply a double- or triple-pole switch in which the moving switch contacts are mechanically linked. In this way both live and neutral or all phases in a three-phase system are disconnected from the supply *(Diagram 82).*

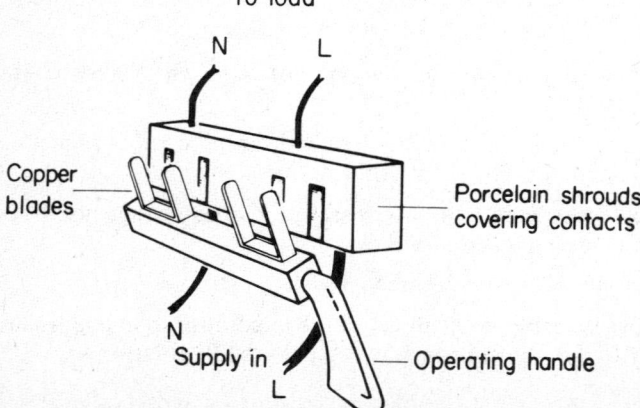

Diagram 82   Internal arrangement of single-phase isolator

Switch fuse

This is an extension of the isolator, in that the load side of the supply is interrupted by a fuse in the phase conductor.

Fuse switch

Fuse switches are used on three-phase systems. Here, the fuse forms part of the moving operating blade.

Consumer unit

The consumer unit is found in nearly all domestic installations. It consists of a double-pole isolator and a distribution board in one assembly.

Circuit breaker

The means of isolation must be double- or triple-pole. Some modern consumer units have a residual current device installed in place of the usual isolator.

Sequence of control

Diagram 83a—c illustrates some typical control sequences.

Points to note (I.E.E. Regulations)

1. Effective and accessible means of isolation must be provided at the origin of the installation to cut off all voltage as may be necessary to prevent danger.
2. When an installation serves two or more detached buildings (i.e. house and detached garage) a means of isolation must be provided in each building.
3. If the purpose of a switch or circuit breaker is not clear, it should be labelled to show which apparatus it controls.

Earthing

Wherever there is a live supply, there is always the risk that metalwork not intended as a conductor (e.g. the metal casing of a cooker) may for one reason or another come into contact with that supply. If this happens two particularly hazardous situations result:

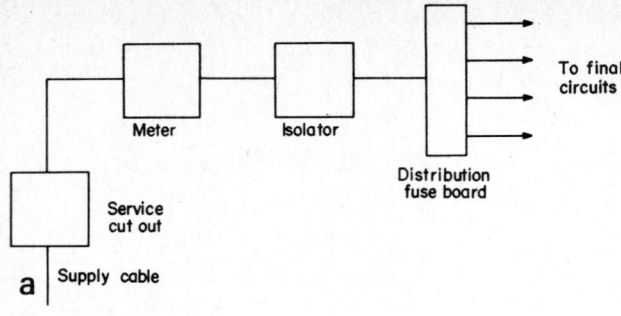
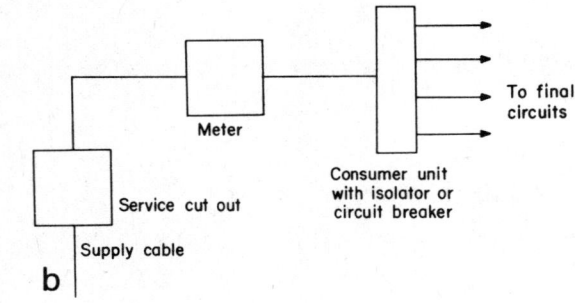
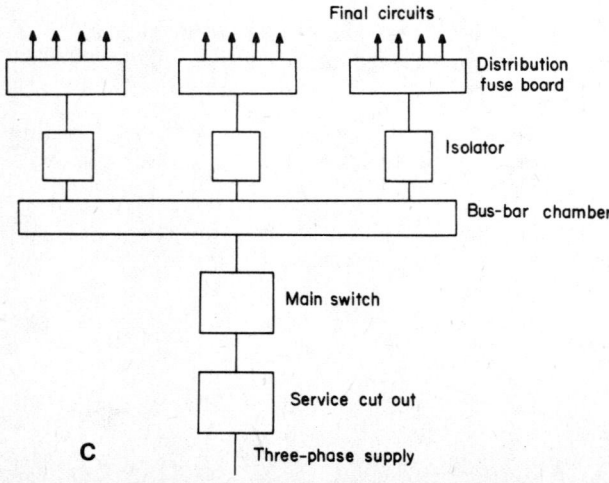

Diagram 83 Typical control sequences

Control and earthing

1. The risk of shock to a person touching the metalwork.
2. The risk of fire due to heavy currents flowing undetected.

The Regulations require that every item of apparatus and every conductor operating in excess of an extra-low voltage shall not be able to cause danger from earth-leakage currents. The methods recommended to prevent this danger are as follows:

1. Enclosing live parts in insulation to provide basic protection against electric shock. This is called basic insulation.
2. The use of double insulation, now increasingly used for portable appliances such as shavers, hair dryers, etc.
3. Earthing exposed metalwork.
4. Isolating metalwork at risk, such that there is no way that it can come into contact with live conductors *or earthed* metalwork.

Items 1 and 2 are usually applied to apparatus designed and manufactured to be free from the danger of earth leakage. Items 3 and 4 are measures which should be taken by the installation electrician.

Earthing of exposed metalwork

Let us consider what should happen under earth-leakage conditions. If metalwork that was not intended as a conductor becomes live (i.e. the metal case of a fire) it presents a risk of shock to anyone touching that metalwork while he is in contact with the general mass of earth. However, if the metalwork were effectively connected to earth by means of a suitable conductor, any connection of the phase conductor to the metalwork would cause enough current to flow to operate the protective device.

In order for this current to flow a complete circuit must exist. *Diagram 84* shows this complete current path. It is called the earth–fault loop path.

This loop path comprises the following parts, starting at the point of fault:

1. The earthed metalwork of the installation (i.e. outer casings of appliances or conduit, etc).
2. The circuit protective conductor.
3. The earthing terminal or block.
4. The earthing conductor.
5. The consumer's earth electrode.

Control and earthing

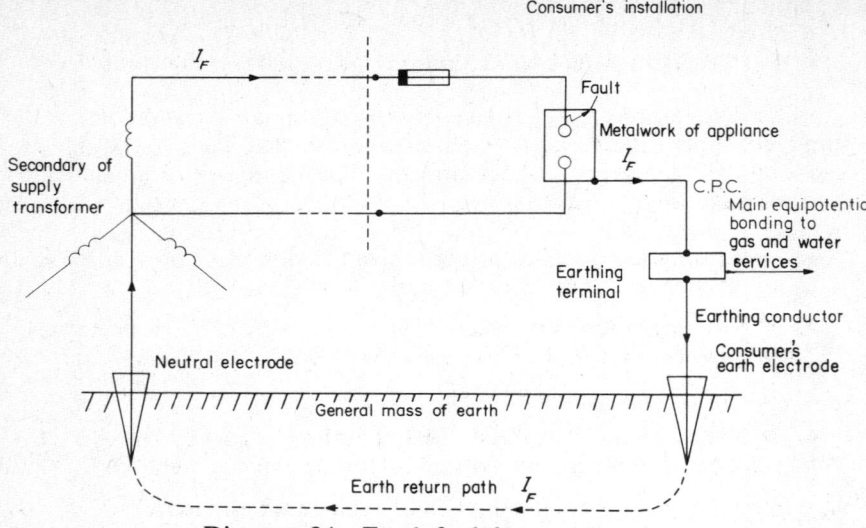

Diagram 84 Earth-fault loop path

6. The general mass of earth or other metallic return conductor.

Note: if the earth return is the metal sheath of the supply cable, a consumer's earth electrode is not usually required.

7. The earthed neutral point of the supply transformer.
8. The phase conductor.

Earth loop impedance

The most important consideration is to ensure that the protective device will operate when a phase-earth fault occurs, and since the value of the fault current I_F depends upon the impedance of the fault circuit, the value of this impedance is of great importance.

In a situation where the earth return path is the metal sheath of the supply cable, the impedance is usually very low and currents of very high values are able to flow, ensuring the immediate operation of the circuit protection. (This assumes of course that the installation has been earthed in accordance with the Regulations.)

Serious problems, however, can arise when the general mass of earth is relied upon to provide the earth return.

Earth as a conductor

Earth or soil itself, whether it is sandy, chalky or rocky, etc., is not a conductor of electricity. It is the moisture in the earth that provides the conducting path. Certain areas, therefore, are better than others in providing a good earth return path. Marshy ground and clay soils for example usually provide a good return path, while gravel and rocky soils are very poor.

Therefore, an accurate measurement of the loop impedance is important to ensure that an installation which relies for its earthing on earth electrodes is adequately protected.

Even when a low reading is obtained, it is *no guarantee* that it will remain at that value. There must have been many installations in a potentially lethal condition during the exceptionally long and dry summer of 1976.

In order that sufficient current will flow to operate the protective device, the Regulations set out a series of tables indicating maximum values of earth loop impedance for each type and rating of protective device.

Hence, an immersion heater circuit protected by a 15 A semi-enclosed fuse to B.S. 3036 would have to have a loop impedance value of 2.7 Ω, i.e. the complete earth circuit must be capable of allowing $240/2.7 = 88.9$ A to flow.

If this value of impedance i.e. 2.7 Ω is not obtainable, then changing the protective device to a type 1 m.c.b. to B.S. 3871 would allow the impedance value to be 4 Ω.

Residual current devices (R.C.D.)

These devices are designed to detect earth-leakage currents almost as soon as they appear. They are fast acting and will interrupt the supply before a serious fault has had time to develop to a dangerous level. They are so sensitive that they are sometimes prone to *nuisance tripping,* in that normal leakage currents that occur in any installation (i.e. cooker rings to casing) can cause the breaker to operate.

Diagram 85a shows a typical residual current device. The circuit diagram of its connection into the system is shown in *Diagram 85b.*

This unit comprises two main coils and a search coil wound on a transformer core, a trip coil and double-pole contacts, and a test circuit.

When the circuit being protected is healthy, the magnetic flux in the core, induced by the current flowing in the live coil, is cancelled out by an equal and opposite flux induced by the return current in the neutral coil. A live-earth fault will at once cause a larger current to flow in the live coil than in the neutral coil. The opposing fluxes are no longer equal and therefore some flux will circulate in the core. This cuts across the conductors of the search coil which will have an e.m.f. induced in it. This e.m.f. will then energize the trip coil and the double-pole contacts will open.

116 *Control and earthing*

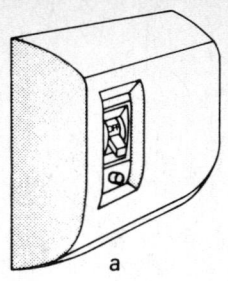

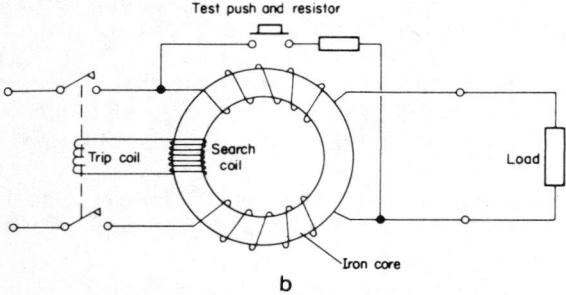

Diagram 85 (a) Residual current device; (b) circuit of residual current device.

The test switch creates an out-of-balance condition which trips the breaker. Its only purpose is to indicate that the breaker is in working order. **It does not check the condition of any part of the earth system.**

Out-of-balance currents as low as 20 mA to 30 mA will be detected, and therefore a person touching unearthed live metalwork would cause the breaker to operate before the lower lethal limit of 50 mA was reached (*Diagram 86*). It is still necessary, however, to ensure that the earth system of an installation is connected to a suitable earth electrode.

Points to note (I.E.E. Regulations)

1. An earthing terminal must be provided adjacent to the consumer's terminals. This is usually in the form of a rectangular metal block with cable entries and screws (*Diagram 87*).

Control and earthing 117

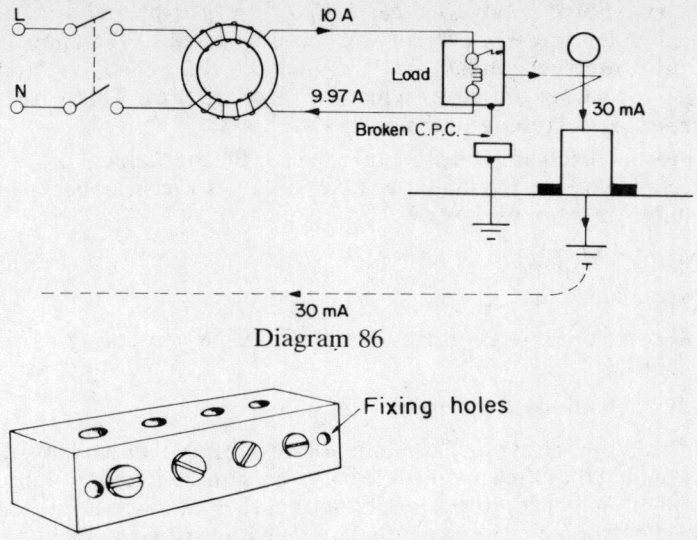

Diagram 86

Diagram 87 Consumer's earth terminal

The earthing conductor from the earth terminal is connected to the cable sheath (is this system is used for earth return) by means of an earth clamp usually of the type shown in *Diagram 88*.

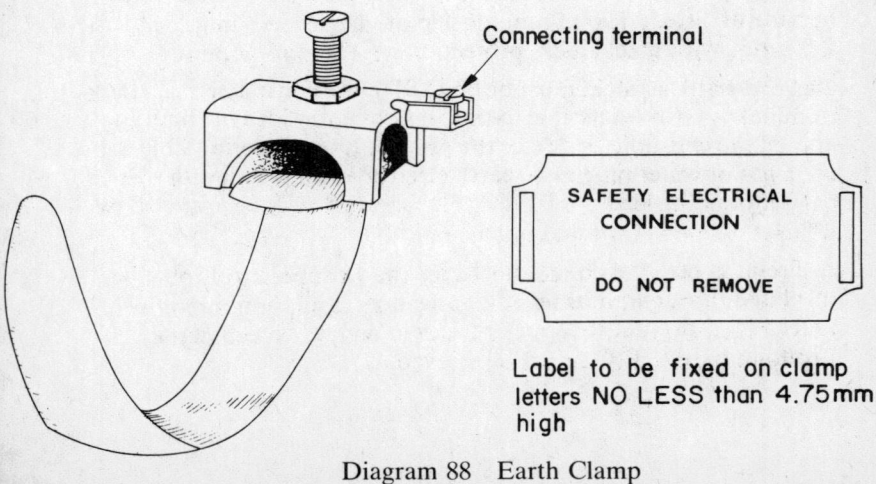

Diagram 88 Earth Clamp

2. All the exposed conductive parts of wiring systems and apparatus not intended to carry current shall be connected to the appropriate circuit protective conductors. This includes things such as metal boxes for socket outlets and metal casings of fires. There are some exceptions to these requirements, as follows:
   (i) Short isolated lengths of metal used for the mechanical protection of cables (conduit used to carry cables overhead between buildings is *not* exempt).
   (ii) Metal cable clips.
   (iii) Metal lamp caps.
   (iv) Metal screws, rivets or nameplates isolated by insulating material.
   (v) Metal chains used to suspend luminaires.
   (vi) Metal luminaires (such as lamp holders) using filament lamps, provided that they are installed above a non-conducting floor and are screened, or positioned so that they cannot be touched by a person able to come into contact with earthed metal.
   (vii) Catenary wires where insulated hangers are used.

   If there is any metalwork in an installation that is likely to come into contact with earthed metal accidentally, then it must either be effectively bonded to, or segregated from, that earthed metal. This includes metal baths, sinks, exposed pipes, radiators, tanks, any structural steelwork that is accessible, and the framework of any mobile equipment such as cranes and lifts which have electrical apparatus fitted. The minimum size of copper bonding lead is 2.5 mm^2, with mechanical protection or 4.0 mm^2 without.

3. Gas and water services must be bonded to the consumer's earthing terminal as near as possible to the point of entry into the building, and on the consumer's side of the service. It is not permissible to use a gas or water pipe as an earth electrode. The minimum size of copper bonding lead is 6.0 mm^2. The bonding may be achieved by using the clamp illustrated in *Diagram 88*.

4. A circuit protective conductor other than copper strip must be insulated throughout its length, and where insulation is removed at terminations (i.e. stripping back twin with earth cable) the resulting bare C.P.C. must be sleeved.

CHAPTER 7

Testing

An essential part of installation work is testing, whether it be on old or new installations or on appliances and fixed apparatus. The sequence of tests, where relevant, which must be carried out on a completed installation is as follows:

1. Continuity of ring final circuit conductors.
2. Continuity of protective conductors.
3. Earth electrode resistance.
4. Insulation resistance.
5. Insulation of site built assemblies.
6. Protection by electrical separation.
7. Protection by barriers or enclosures provided during erection.
8. Insulation of non-conducting floors and walls.
9. Polarity.
10. Earth fault loop impedance.
11. Operation of residual current and fault voltage devices.

Continuity of ring final circuit conductors

This test ensures that the Regulations are complied with, in that it checks that each conductor forming the ring starts at the distribution board, loops into each socket or joint box on the ring, and returns to the same way in the distribution board. *Diagram 89* illustrates this.

Procedure

(i) Disconnect the supply and remove circuit fuses.

(ii) Disconnect each pair of ring conductors in turn.

(iii) Test between each pair; note this resistance R ohms.

120 Testing

(iv) Complete the ring connection and test between this connection and the corresponding terminal of an outlet nearest to the mid-point of the ring. Note this resistance Rr.

(v) Deduct the resistance of the test leads R_L from Rr. The resulting value should be approximately ¼ of R.
i.e. $Rr - R_L = \quad R/4$

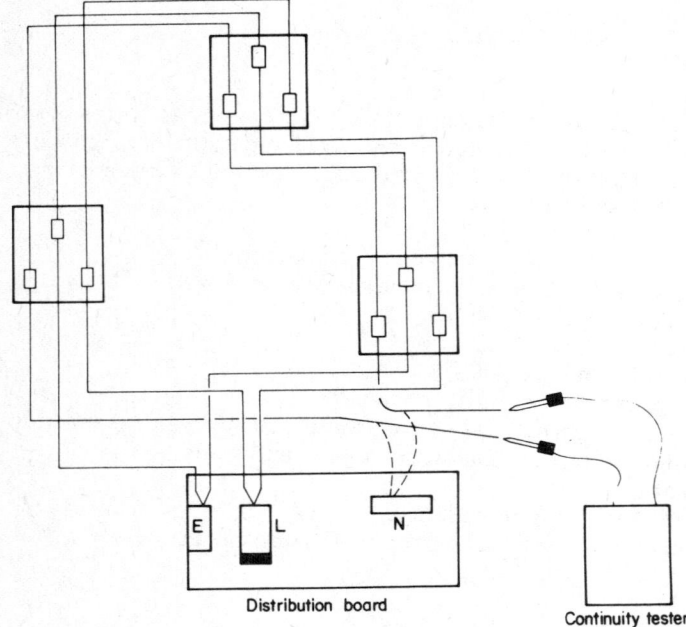

Diagram 89   Ring continuity test

Polarity

This test is done to ensure that all fuses and single-pole devices (switches) are connected into the live conductor, that the centre contact of an Edison-type screw lampholder is connected to the live conductor and that all plugs and sockets are correctly wired.

There are two types of test, the live circuit test and the dead circuit test. The live test requires the use of a voltmeter or an approved test lamp. This method, however, has the disadvantage that work is carried out on live equipment. The dead circuit test is more widely used. The method of testing is shown in *Diagrams 90* and *91*.

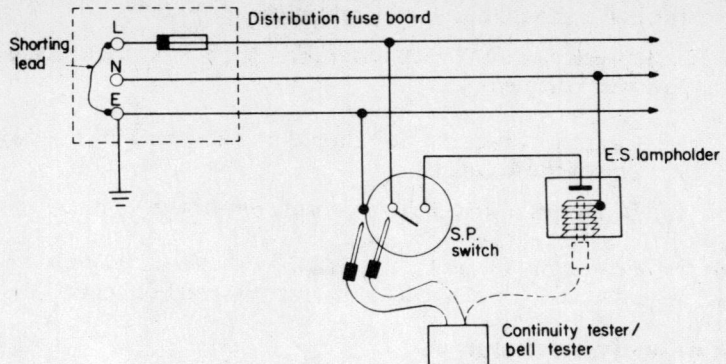

Diagram 90   Circuit polarity (lighting)

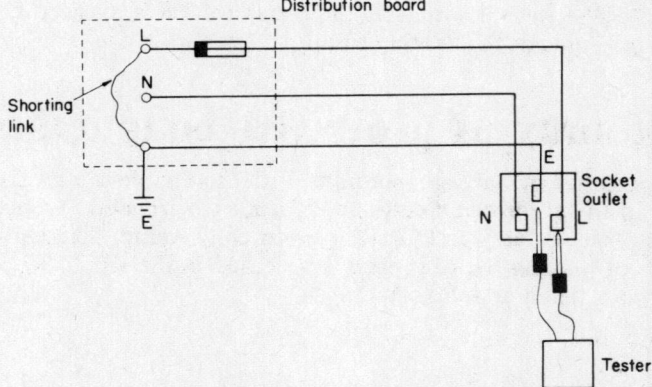

Diagram 91   Circuit polarity (sockets)

Circuit polarity (lighting)
Procedure
- (i) Open the main switch and remove the consumer's main fuses (if any), ensuring that all fuses are in the distribution board.
- (ii) Link the live (on installation side of main switch) to consumer's earthing terminal.
- (iii) Remove all lamps.

(iv) All switches should be in the *off* position.

(v) Test between earth and each terminal of the S.P. switch. Correct polarity is indicated by:

    (a) Continuity tester. There should be a zero reading on only *one* of the terminals.

    (b) Bell tester. The bell rings on only *one* of the terminals.

Note: To test the polarity of the E.S. lampholder shown in *Diagram 90*, the switch would have to be closed after its polarity had been checked.

Circuit polarity (sockets)
Procedure

(i) Remove all appliances.

(ii) Test between (L) right-hand pin and the earth. Correct polarity is indicated as for lighting circuit.

Continuity of protective conductors

The importance of having a sound C.P.C., to enable the protection to operate in the event of an earth fault, has been stressed in Chapter 6. This test ensures that the C.P.C. is electrically sound. There are several methods of making the test; each is specified in the I.E.E. Regulations, the methods being as follows:

1. a.c. test.
2. d.c. test.

Regardless of the method by which the test is made, the principle remains the same. The continuity of each C.P.C. is measured between the consumer's earthing terminal and the far end of the C.P.C.

A.c. test procedure

(i) The final circuit is disconnected from the supply.

(ii) Test equipment is connected between the consumer's earth terminal and the far end of the C.P.C.

(iii) A current of 1.5 times the circuit current is then injected at 50 V, 50 Hz (current need not exceed 25 A).

Diagram 92 shows the method of connection.

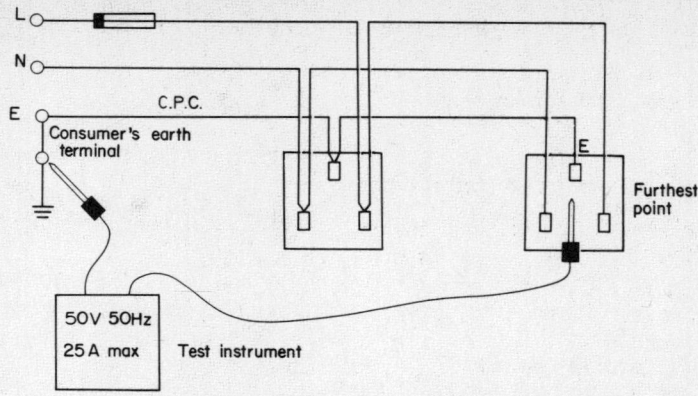

Diagram 92   Earth continuity test

D.c. test

This is the most commonly used method, the test instrument being a continuity tester. Care must be taken to ensure that no inductor is included throughout the length of the C.P.C.

Confirming the soundness of the C.P.C. is not sufficient. As fault currents have to flow around the whole earth-fault loop path, a measurement of the earth loop impedance must be taken.

Earth electrode resistance

It is often required to measure the resistance of an earth electrode, as it forms part of the total loop impedance. The Regulations detail the method to be used. It is also described further in volume 3.

Loop impedance measurement

There are two recognized tests for this, the live-earth test and the neutral-earth test.

Live-earth test

The procedure for this test is to connect the live to the earth through a resistor of known value. The resulting current, together with the supply voltage enables the impedance of the loop path to be calculated. In practice, the live-earth loop tester gives a direct reading in ohms.

Neutral-earth test

This test involves the injection of current into the neutral-earth loop. However, the restriction on its use with certain earthing systems

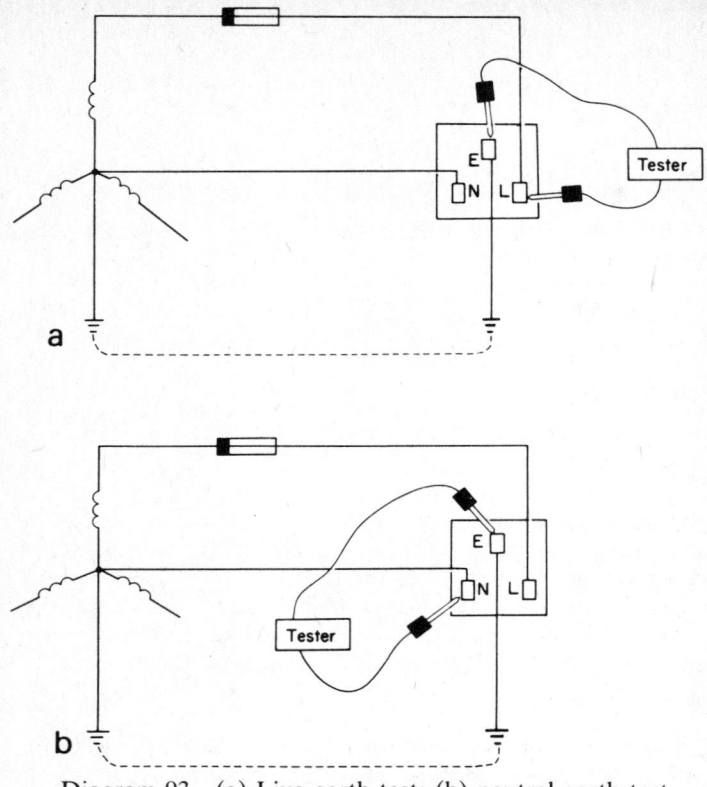

Diagram 93   (a) Live-earth test; (b) neutral-earth test

(TN-C-S) together with the adjustments to measurements necessary in order to compensate for the normal neutral currents, make this test unpopular.

The test instruments for both tests are connected as shown in *Diagram 93*.

Insulation resistance tests

In order to ensure that the cable insulation is sound and that no live or neutral earth faults exist on an installation, an insulation resistance test is carried out. Clearly, the higher the resistance of the insulation between conductors, and between conductors and earth, the better. The Regulations recommend a minimum value of 1 megohm (MΩ).

Phase neutral-earth test

This is carried out with an insulation resistance tester (megger) which will supply a voltage of not less than twice the normal supply voltage. It need not exceed 500 V.

Procedure
- (i) Isolate supply as near to main intake position as is possible.
- (ii) Ensure that all final circuit fuses are in place, and that all switches are closed.
- (iii) Link all poles of the supply together.
- (iv) The test is made between the linked poles and earth.

Between-poles test

The same instrument is used.

Procedure:
- (i) Isolate the supply.
- (ii) Remove all lamps from holders and all appliances from sockets. If this is not possible, ensure that the switch is *off*.
- (iii) Test between poles.

Diagrams 94 and *95* give details of both tests.

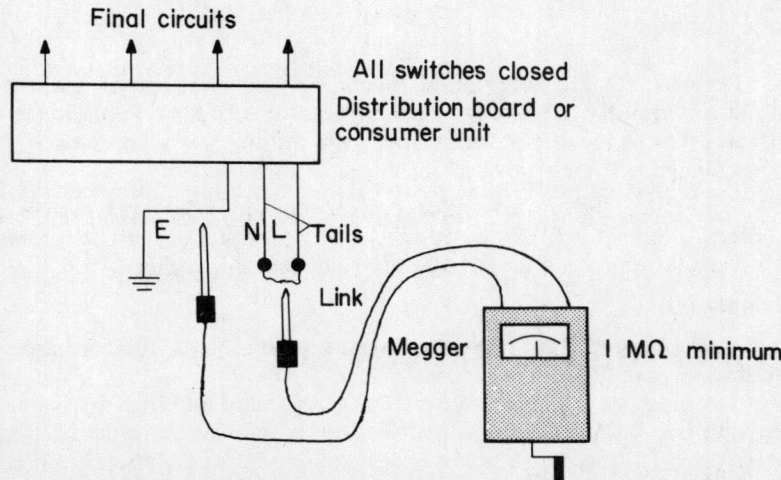

Diagram 94   Insulation resistance to earth

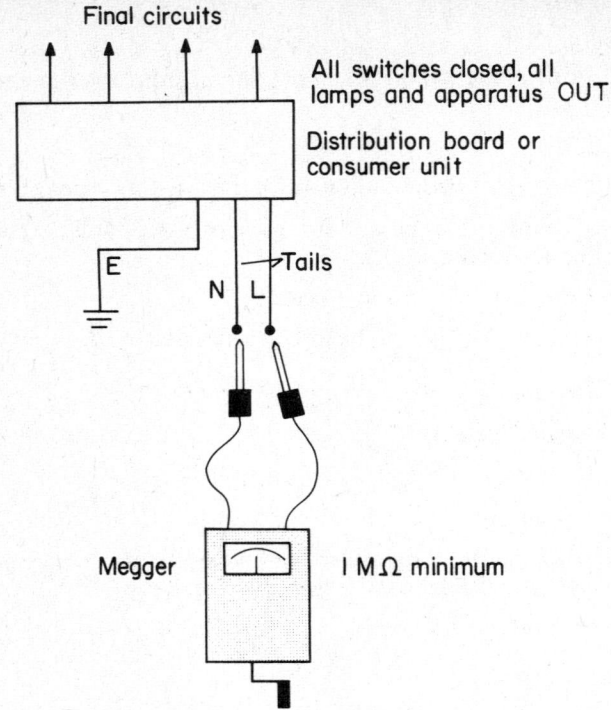

Diagram 95   Between-poles test

It is interesting to note that an insulation resistance test on a whole installation may give a reading below 1 MΩ, and yet tests on individual circuits, made to locate the faulty one, show readings at or in excess of 1 MΩ. *Diagram 96* helps to explain this situation.

As the insulation resistance consists of countless resistances in parallel and since the live and neutral are common to both circuits, then all the insulation resistances of all the circuits are in parallel.

Example 7.1

A between-poles insulation resistance test on individual circuits of an installation gave the following results.

Lighting — 250 MΩ

Ring main — 25 MΩ

Water heater — 4 MΩ

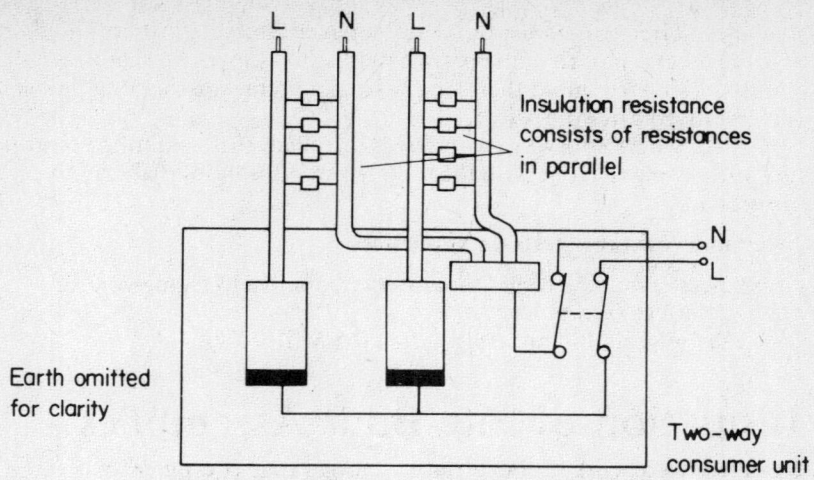

Diagram 96   Insulation resistances in parallel

Cooker — 12.5 MΩ

Radial circuit — 1.14 MΩ

What would be the reading taken at the supply terminals, with all circuit fuses in place?

Insulation resistances are in parallel, therefore:

$$\frac{1}{R_t} = \frac{1}{250} + \frac{1}{25} + \frac{1}{4} + \frac{1}{12.5} + \frac{1}{1.14}$$

$$\frac{1}{R_t} = 0.004 + 0.04 + 0.25 + 0.08 + 0.877$$

$$\frac{1}{R_t} = 1.25$$

$$\therefore R_t = \frac{1}{1.25}$$

$$= 0.8 \text{ M}\Omega \text{ which is not acceptable}$$

Important note

Many modern appliances and accessories have electronic control gadgets inside them such as dimmer switches, and timers on cookers and washing machines. Great care must be taken, when conducting insulation resistance tests, not to damage such items. It is suggested that they should be shorted out at their supply intake point; the shorting link is then removed after the test.

Also, where two-way switching is installed, the insulation resistance test must be carried out with the switches in *both* of their alternative positions.

Insulation resistance tests on appliances

Appliances should be tested separately, from live to framework or casing, and between poles.

The value should not be less than 0.5 MΩ.

Insulation of Site Built Assemblies

It sometimes occurs in industrial situations that equipment has to be erected and insulated on site and it is therefore important to test this insulation to the British Standards for similar factory built equipments.

Electrical Separation

Here the test is usually an inspection to ensure that certain circuits are separated from one another.

Protection against direct contact by barriers etc.

If direct contact with live parts is to be prevented by barriers or enclosures, it must be verified that they conform with B.S. 5490 IP2X or IP4X.

IP2X is a standard test, applied with a standard test finger, to ensure that when inserted, the finger is protected by barriers from touching live parts.

IP4X is similar, but here, a 1 mm diameter rigid steel wire is used instead of the standard finger.

Insulation of non-conducting floors or walls

If protection from shock is to be afforded by non-conducting floors or walls, the insulation resistance between the floors and walls, and the main protective conductor must be measured at at least three points on each relevant surface, one of which must be between 1 m and 1.2 m from any extraneous conductive part in the area. The reading obtained should not be less than 50 kΩ for a supply voltage less than 500 V or 100 kΩ over 500 V, but not exceeding low voltage.

Operation of residual current devices

In this case a fault is simulated to ensure the effectiveness of the device. The procedure is outlined in the I.E.E. Regulations and is discussed further in Volume 3.

Inspection

So far, we have dealt with particular tests carried out with specially designed instruments. However, another important aspect in the checking of an installation is the *visual inspection*. The following is a list of some of the more important things to look for when checking that an installation complies with the I.E.E. Regulations and /or is good working practice.

Mains intake position

Correct means of isolation.

Rating of 'tails' is not less than the main fuse rating.

A consumer's earthing terminal has been provided and correct sizes of C.P.C., earthing conductor, and equipotential bonding conductor have been used.

Neutrals in the distribution board are wired in the same order as the live conductors.

A list of circuit numbers and functions (i.e. No. 3—Cooker) is on, or in, or adjacent to the distribution board.

Distribution boards without backs are not fitted to combustible surfaces.

Ensure that the means of isolation is accessible, in the event of an emergency.

All bare C.P.C.s are sleeved in green/yellow sleeving.

Ring and radial circuits

The number of non-fused spurs does not exceed the number of points on the ring.

Cable is correct size.

Metal boxes are adequately bushed and are of sufficient depth.

All bare C.P.C.s are sleeved in green/yellow sleeving.

Sockets are not less than 150 mm above the floor or working surface.

Metal boxes with adjustable lugs have an earthing tail from the socket earthing terminal to the terminal on the box.

Sockets are not installed in bathrooms or within 2.5 m of any shower unit in a room other than a bathroom.

Lighting circuits

Switches are accessible.

S.P. switches are connected in the live conductor.

Switch wires are identifiable as live conductors.

C.P.C. is wired to earthing terminal of switch box or ceiling rose.

All bare C.P.C.s are sleeved in green/yellow sleeving.

Switches are not accessible to a person using a fixed bath or shower.

Correct current rating has been used.

Correct type used for fluorescent lighting or de-rated to suit.

Ceiling roses have a shrouded live terminal and are suitable for the mass suspended.

Lampholders in damp situations or accessible to persons in contact with earthed metal or within 2.5 m of a bath or shower should be fitted with 'Home Office skirts' (*Diagram 97*).

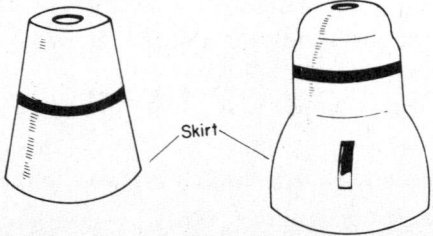

Diagram 97   Lampholders with Home Office skirts

Sundry items
Cooker control unit is within 2 m of cooker, and readily accessible.

Shaver units in bathrooms are suitable for that use.

Joint boxes are readily accessible.

All cables are of correct size and are protected from mechanical damage.

All circuit protection is of correct size and type.

Cables pass through joists 50 mm from top or are protected with a short length of conduit if the joist is 'notched' at the top.

Conduit, trunking and traywork
Space factor is not exceeded.

Conduit ends are reamed and bushed.

Drainage holes are provided (conduit).

Securely fixed.

Corrosion-resistant in damp situations.

Flexible conduit not used as C.P.C.

Trunking lids are securely fixed.

Cables of different categories are segregated.

All phases and neutrals are in the same conduit or trunking.

Barriers provided to prevent spread of fire.

Metal traywork earthed if only supporting P.V.C.-sheathed cables.

Cables
Correct rating and type.

Adequately supported to prevent strain.

Not too much insulation removed.

All joints are electrically and mechanically sound.

Adequately protected against mechanical damage.

General
Ensure that all disturbance to walls, floors and ceilings is made good.

All accessories are of an approved type.

All accessories and fittings are securely fixed and lined up correctly.

Instruments

Instruments play an important part in installation work enabling the measurement of the current, voltage, resistance, power and power factor.

The basic ammeter and voltmeter work on either the moving-iron or moving-coil principle.

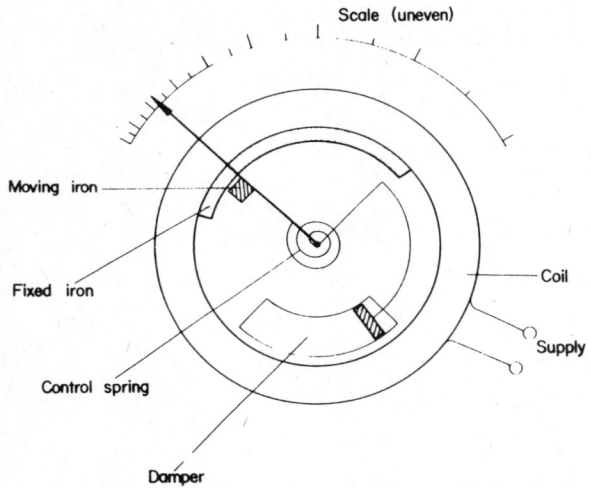

Diagram 98   Moving-iron instrument

The moving-iron instrument (repulsion type)

Diagram 98 illustrates a moving-iron instrument of the repulsion type. It comprises a coil, with a fixed iron, a pointer with an iron vane attached (moving iron) and a damping device inside it.

When a supply, either a.c. or d.c., is applied to the coil, both fixed and moving irons are magnetized to the same polarity and will therefore repel each other.

The design of the irons ensures that the repulsion is always in the same direction.

The damper ensures a slow and even movement of the pointer. It consists of a cylinder closed at one end and a light piston inside it. The pointer, which is attached to the piston, is slowed down by the air pressure which builds up in the cylinder, resisting the movement of the piston. A spring returns the pointer to zero when the supply is removed.

As the amount of movement depends on the square of the supply current, a small current produces a small movement and a large current a larger movement. Hence the scale tends to be cramped at the lower values of the current.

The moving-coil instrument

These work on the the motor principle of a current-carrying coil in a magnetic field. *Diagram 99a* and *99b* shows two variations of this type of instrument.

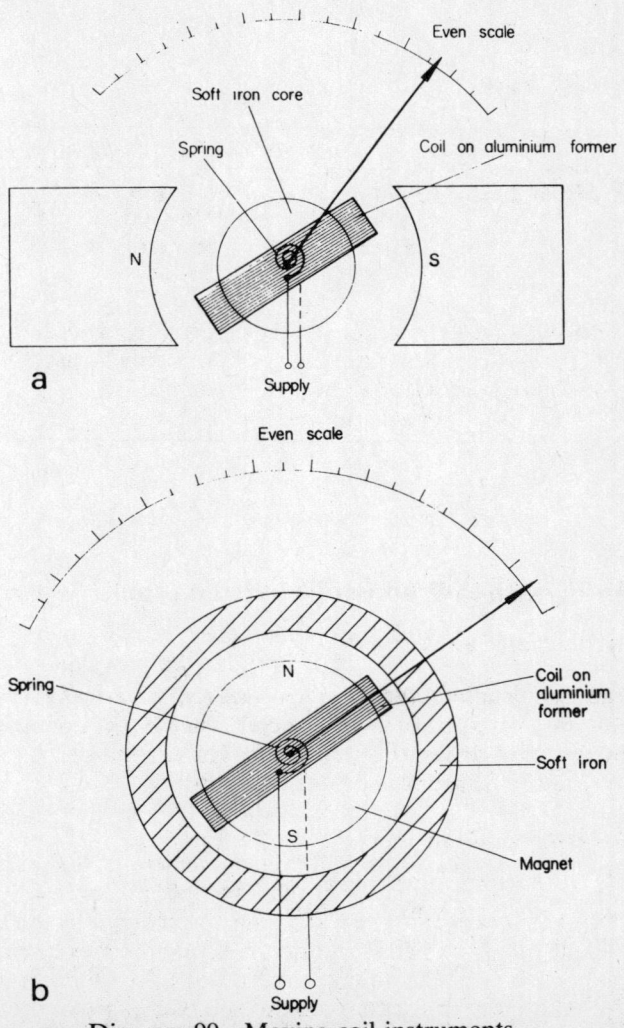

Diagram 99   Moving-coil instruments

The basic components of both the systems are a magnetic field, a core or shell of soft iron and a coil wound on an aluminium former; connection to the coil is made via the control springs.

Damping is achieved by eddy currents in the aluminium coil former. These currents cause small magnetic fields to flow which interact with the main field and cause the movement of the coil to slow down.

Comparison of types

	Advantages	*Disadvantages*
Moving iron	Cheap, strong, can be used on a.c. and d.c.	Uneven scale, affected by heat and stray magnetic fields
Moving coil	Even scale, very accurate, unaffected by stray magnetic fields	Fragile, expensive, can only be used on d.c.

Installation testing instruments

Insulation resistance tester

This instrument, usually referred to as a *megger,* is basically a moving-coil meter with a voltage and a current coil. The voltage coil is supplied directly by the hand-driven d.c. generator. The current coil is fed from the same source but through the external circuit.

Continuity tester or ohmmeter

This works on the same principle as the megger except that smaller values of resistance can be measured. Some meggers incorporate both instruments in one, with a selection switch for the ohms or the megohms scale. Electronically operated testers powered by 9 V batteries, giving up to 1000 V output are also available.

Diagrams 100, 101 and *102* show some typical test instruments.

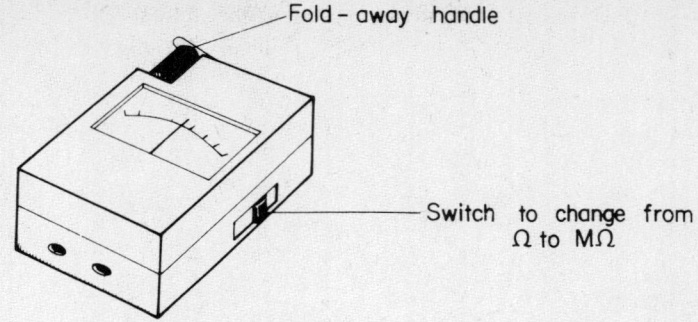

Diagram 100   500 V insulation resistance/continuity tester (megger)

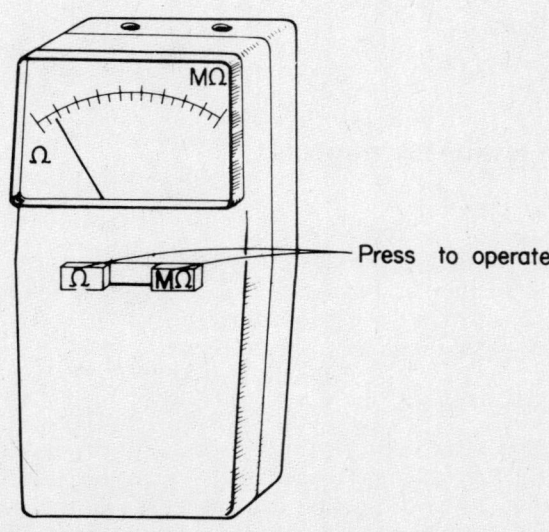

Diagram 101   Electronic type megger

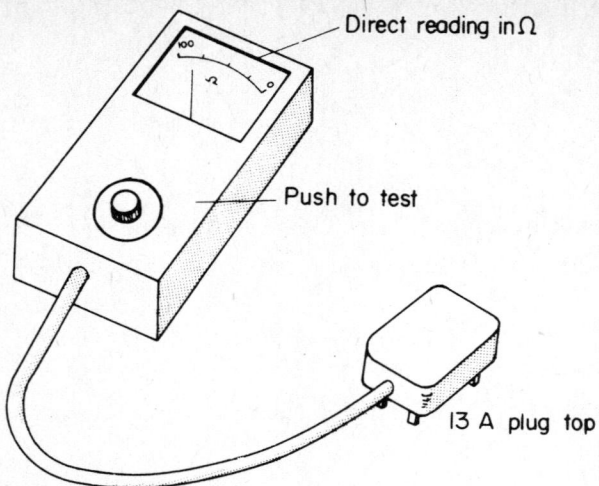

Diagram 102 Electronic phase-earth loop tester

CHAPTER 8

Basic Electronics Technology

Most of us in the world of electrical installation work are familiar with values such as 240 V, 3 kW or 60 A. We tend to view the terms and quantities used in electronics with a certain unease, but the relationship between ohms, volts and amperes in the world of electronics is no different from that in installation work.

In order to begin to feel more at home with electronics, we first look at some of the many components used.

Electronics components

Resistors

The ohmic value of 240 V appliances rated at 60 W, 1000 W, 3 kW etc. should be familiar to us by now, but the values of resistors used in electronics are many and varied. In order to identify readily all the different values, a colour code is used; the same code is used for capacitor values.

Resistor (or capacitor) colour code

Colour	Value
Black	0
Brown	1
Red	2
Orange	3
Yellow	4
Green	5
Blue	6
Violet	7
Silver	8
White	9

Tolerance colour code

Colour	Percentage
Brown	1%
Red	2%
Gold	5%
Silver	10%
None	20%

138 *Basic electronics technology*

Each resistor carries a series of coloured bands to indicate its value and tolerance:

3 bands for a resistor with tolerance of 20%.
4 bands for a resistor with tolerance of between 10% and 2%.
5 bands for a resistor with tolerance of 1%.

These bands are interpreted as shown in *Diagram 103*. *Diagram 104* shows two examples of how to 'read' the colour coding to ascertain the value and tolerance.

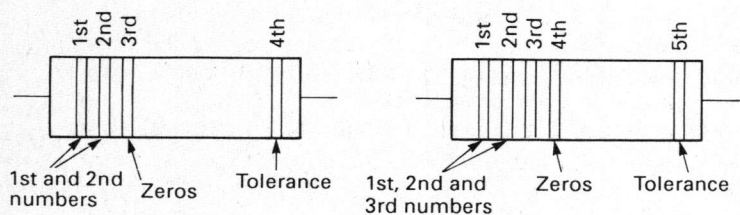

Diagram 103 Coding bands

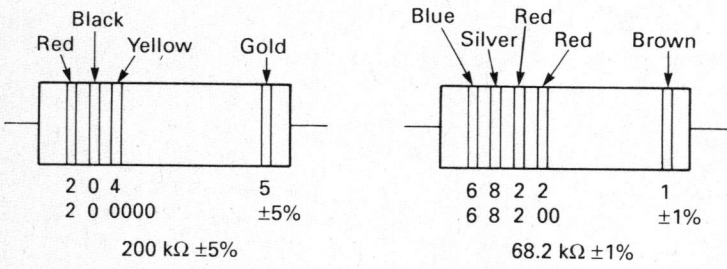

Diagram 104 Colour values decoded

Sometimes a resistor code will use numbers and letters rather than colours. The letters used are as follows:

R, K, M indicate multipliers of 1 Ω, 1000 Ω, 1000 000 Ω respectively
F, G, J, K, M indicate tolerances of 1%, 2%, 5%, 10%, 20% respectively.

The R, K, M code can also be used for decimal points, e.g.

3.3 kΩ may be shown as 3K3 Ω.
0.2 Ω may be shown as R2 Ω.

Other examples are:

2K2G   indicates a value of 2.2 kΩ (2200 Ω) ± 2%.
6M8J   indicates a value of 6.8 MΩ (6800 000 Ω) ± 5%.
33KF   indicates a value of 33.0 kΩ (33 000 Ω) ± 1%.
470RM indicates a value of 470.0 Ω ± 20%

Resistor types

There are three types of resistor in common use: carbon, wire-wound and carbon pre-set or variable.

The wire-wound type is usually chosen where high voltage is present; it is also more accurate than the carbon variety. The pre-set type, shown in *Diagram 105*, is used in the simple metal detectors used by electricians and to adjust EXIT and ENTRY times in alarm panels.

Diagram 105   Pre-set resistor

Capacitors

There are a number of ways of marking a capacitor with its value. The most common, apart from actually writing the value on the capacitor, is to use the same colour code as for resistors (*Diagram 106*). Remember, small capacitors are usually in the picofarad (pF) range (1 nanofarad (nF) = 1000 pF).

One other method commonly used is the three-digit method, in which the third digit gives the number of zeros that follow the first two digits, to give the value in pF: e.g. 104 is the code for a capacitor of 100 000 pF or 100 nF.

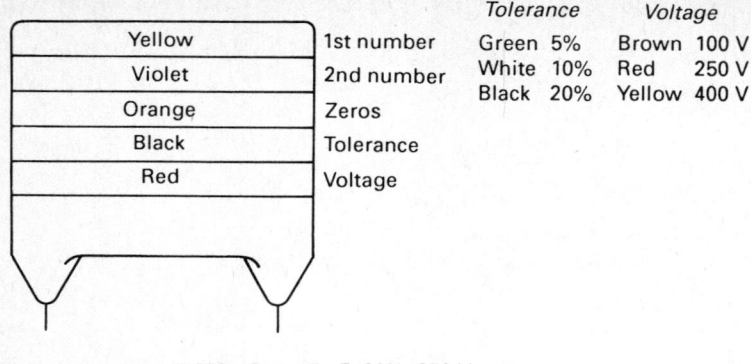

Diagram 106 Capacitor colour coding

Capacitor types

The many types of capacitor in use range from waxed paper and foil, electrolytic, polyester, mica and ceramic, to air—all of which have different applications depending on frequency, voltage, supply (a.c. or d.c.), losses etc.

Inductors and transformers

Inductors and transformers have already been discussed in Chapters 1 and 4. The types used in electronics work on the same principles. Probably the main difference is the use of ferrite as a core for inductors that are used to tune for radio frequencies. This is a much more efficient material than iron as the hysteresis losses are much less.

Semi-conductors

These devices are neither strictly conductors nor insulators but, under certain circumstances, can become either.

Silicon is the most common semi-conductor material, and the addition of impurities such as aluminium or arsenic creates the circumstances under which it will conduct or insulate. For example, if we add aluminium to a sample of silicon it becomes what is known as a 'P' type material; adding arsenic to the sample makes it an 'N' type material.

The junction diode

If we now take a sample of 'P' type and a sample of 'N' type silicon and join them together we will have a 'junction diode'. By connecting a positive charge to the 'P' type sample and a negative charge to the 'N' type, the whole assembly will act as a conductor. Reversing the connections will result in the arrangement acting as an insulator (*Diagram 107*). Hence it is used commonly as a means of rectification.

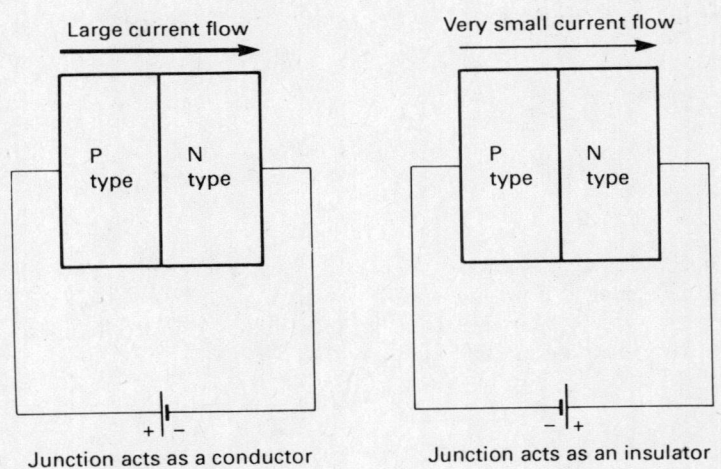

Diagram 107 Semi-conductor

This arrangement is the basis for all other semi-conductor devices, as they are no more than variations on a series of P–N junctions. The following is a list of the more common types.

Transistor—In its basic form this is used as an amplifier in that a small voltage applied to it will allow large currents to flow through it.

Zener diode—This is used in circuits to give voltage control.

Thyristor—This device is a rectifier which can control the amount of current flowing and is used as a means of speed control for small motors.

Triac—Triacs are often used 'back-to-back' to provide a smoother and more efficient thyristor effect. They will often be found in dimmer switches.

Diac—Used in conjunction with triacs as a triggering device.

Thermistor—Used as a means of sensing temperature change. Commonly found embedded in motor windings to detect overheating.

L.E.D.—Light emitting diode. This is simply a semi-conductor signal lamp.

L.S.D.—Light sensitive diode, used to activate a circuit in response to light.

Heat sinks

In many instances, semi-conductors will become hot when in operation and would be damaged if this heat were not dissipated. Embedding the device in the centre of a large plate or series of plates helps to dissipate the heat to the surrounding air (*Diagram 108*).

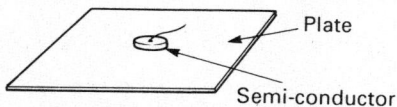

Diagram 108   Heat sink

Electronics diagrams

Three main types of diagram are used:
1. **Block diagrams** (*Diagram 109*) are used to give a general indication of a complete system.
2. **Circuit diagrams** (*Diagram 110*), as in installation work, indicate how a circuit *works*.
3. **Layout diagrams** (*Diagram 111*) are similar to wiring diagrams in installation work: they show how a circuit should be *wired* and components are shown in their correct locations.

Basic electronics technology 143

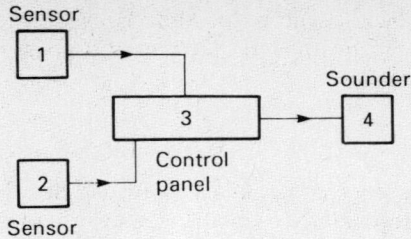

Diagram 109 Block diagram of an intruder alarm system

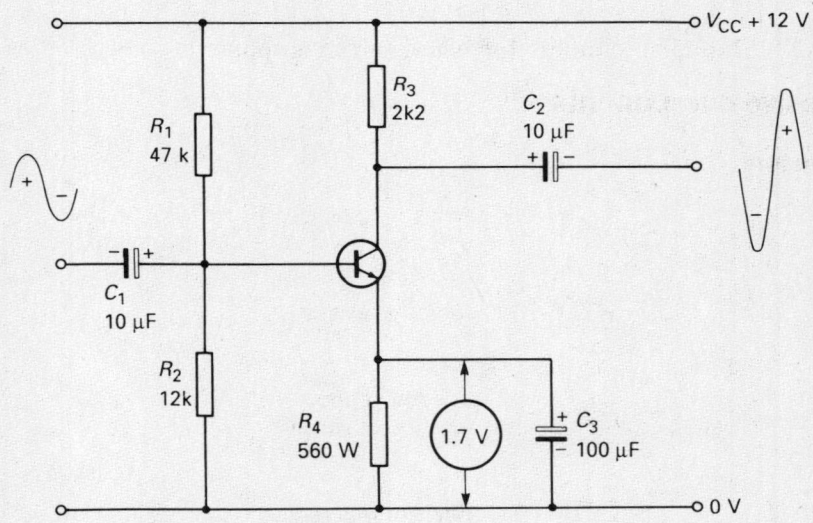

Diagram 110 Simple transistor amplifier circuit

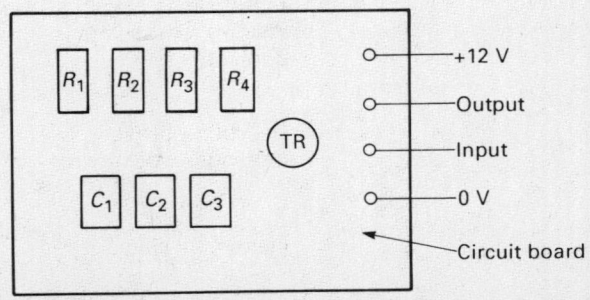

Diagram 111 Layout diagram of an amplifier circuit

144 *Basic electronics technology*

Note: Cross-reference should be possible between circuit and layout diagrams by means of pin numbers of a positional reference system.

Symbols

All diagrams should show components to British Standard 3939. A selection of the more common symbols is shown below. These extracts from B.S.3939: Part 2: 1985 are reproduced with the permission of BSI; complete copies of the standard can be obtained by post from B.S.I. Sales, Linford Wood, Milton Keynes MK14 6LE.

British Standard symbols for electronics components

Passive components

Resistors

04-01-01	 Resistor, general symbol
04-01-03	 Variable resistor
04-01-04	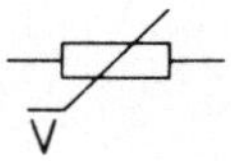 Voltage dependent resistor, or varistor
04-01-05	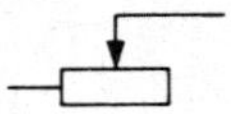 Resistor with sliding contact
04-01-07	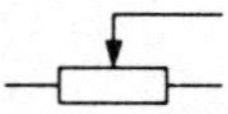 Potentiometer (voltage divider) with sliding contact
04-01-09	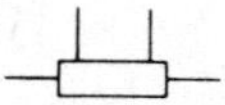 Resistor with fixed tappings, two shown
04-01-10	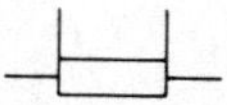 Instrument shunt
04-01-12	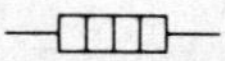 Heating element

Basic electronics technology 145

Capacitors

04-02-01		Capacitor, general symbol
04-02-03		Feed-through capacitor
04-02-05		Polarized capacitor, for example, electrolytic
04-02-07		Variable capacitor
04-02-09		Capacitor with pre-set adjustment
04-02-16		Voltage dependent polarized capacitor, for example, semi-conductor capacitor

Inductors

04-03-01		Inductor, coil, winding, choke
04-03-03		Inductor with magnetic core
04-03-04		Inductor with gap in magnetic core
04-03-05		Continuously variable inductor with magnetic core

04-03-06 Inductor with tappings, two shown

04-03-10 Ferrite bead, shown on a conductor

Ferrite cores and magnetic storage

04-04-01 Ferrite core

04-06-01 Ferrite core matrix with x and y windings and a read-out winding

Semiconductors

Semiconductor diodes

05-03-01 Semiconductor diode, general symbol

05-03-02 Light emitting diode, general symbol

05-03-04 Variable capacitance diode (varactor)

Basic electronics technology 147

05-03-05		Tunnel diode
05-03-06		Breakdown diode, unidirectional. Voltage regulator diode. Esaki-diode

Thyristors

05-04-04		Triode thyristor, type unspecified *Note –This symbol is used to represent a reverse blocking triode thryristor, if it is not necessary to specify the type of gate*
05-04-05		Reverse blocking triode thyristor, N-gate (anode-side controlled)
05-04-06		Reverse blocking triode thyristor, P-gate (cathode-side controlled)
05-04-08		Turn-off triode thyristor, N-gate (anode-side controlled)
05-04-11		Bidirectional triode thyristor. Triac

Transistors

05-05-01		PNP transistor
05-05-02		NPN transistor with collector connected to the envelope

148 *Basic electronics technology*

05-05-04		Unijunction transistor with P-type base
05-05-05		Unijunction transistor with N-type base
05-05-09		Junction field effect transistor with N-type channel
05-05-10		Junction field effect transistor with P-type channel
05-05-12		IGFET enhancement type, single gate, N-type channel without substrate connection
05-05-13		IGFET enhancement type, single gate, P-type channel with substrate connection brought out
05-05-15		IGFET, depletion type, single gate, N-type channel without substrate connection
05-05-16		IGFET, depletion type, single gate, P-type channel without substrate connection
05-05-17		IGFET, depletion type with two gates, N-type channel with substrate connection brought out

Basic electronics technology 149

Electronics assembly

Unlike installation circuits, electronic circuitry is almost entirely constructed using soldered joints. The formation of such joints is critical to ensure healthy circuit performance.

Soldering

A soldered joint comprises the surfaces to be joined and a material (solder) which is an alloy of tin and lead, melted on to the surfaces. To aid the soldering process a flux is used and usually this is incorporated in the solder.

Cleanliness is vital to ensure a good soldered joint—cleanliness not only of the surfaces to be joined but also of the 'bit' of the soldering iron.

Remember, when soldering, that *too much* solder will make a poor joint and that the iron should be at the correct temperature. Many modern irons have built-in temperature controls. *Diagram 112* shows some effects of soldering faults.

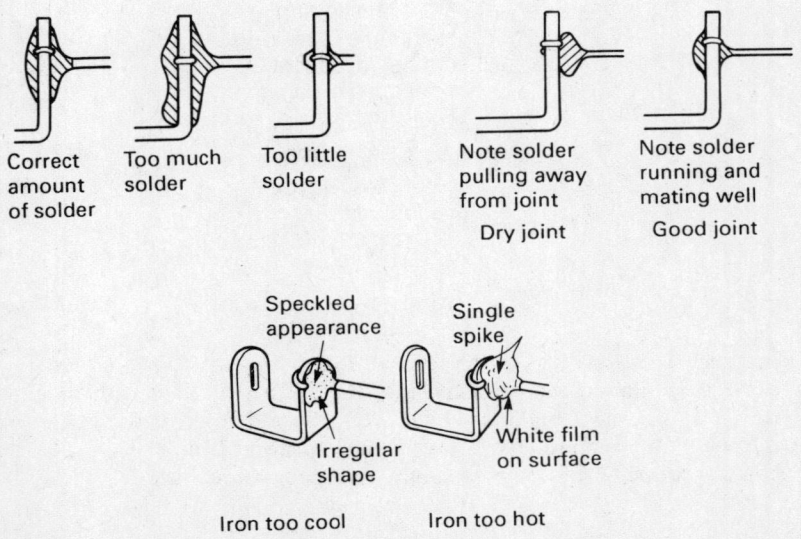

Diagram 112   Soldered Joints

Heat shunts

As heat can damage electronics components, it is important to ensure that too much heat does not reach the component during the soldering process. Heat shunts are used for this purpose and can simply consist of a pair of pliers or a crocodile clip attached to the component lead. This helps to dissipate the heat before it can reach the component itself (*Diagram 113*).

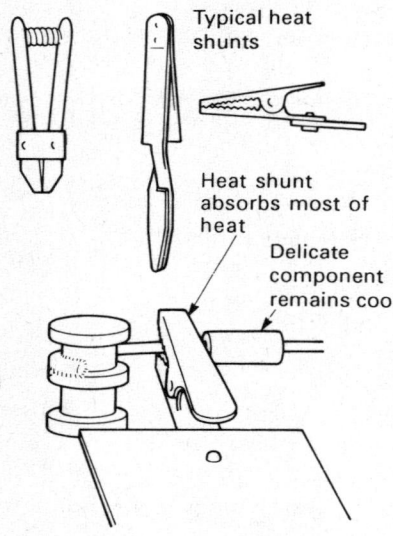

Diagram 113   Heat shunts

De-soldering

It may be necessary on occasion to remove a component from a circuit. The process is similar to that of soldering: the joint is flooded with new hot solder and, as a heat balance is achieved, the old solder softens. When the whole lot is fluid, the solder is sucked away from the joint using either a copper braid or a specially designed 'solder-sucker'.

Miscellaneous Questions on Part I

VOLUME 1, CHAPTER 1

1. What are the I.E.E. Wiring Regulations?

2. What does the Health and Safety at Work Act 1974 say?

3. What is an inclined plane? Name two types of tool using the inclined plane principle.

4. Give examples of first- and second-order levers.

5. Calculate the effort required to lift the load in *Diagram 114*.

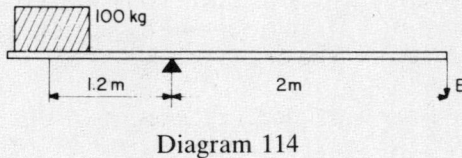

Diagram 114

6. What precautions should be taken regarding the positioning and securing of ladders?

7. What is meant by a *ladder scaffold?* What is the maximum span length of such a scaffold?

8. Above what height should toe boards and hand rails be used?

9. Name the three elements, the combination of which can result in fire. Upon what principle is the extinguishing of fire based?

10. Fires are classed as A, B or C. What kind of fires come under each class?

11. What types of extinguishers are used for each kind of fire? What colour code is used to identify extinguishers?

12. What precautions must be taken when soldering?

152 *Miscellaneous questions on Part I*

13. You discover a person unconscious and in contact with live electrical equipment. Describe step by step the action you would take to aid him or her.

VOLUME 1, CHAPTER 2

14. What is an *electron?* Why are some electrons free to wander at random in a material?

15. Explain why some materials are good conductors, and others are good insulators. Give three examples of each.

16. (a) Define the *coulomb*. (b) How long will it take for 39 C to pass in circuit carrying a current of 13 A?

17. (a) State Ohm's Law. (b) Calculate the resistance of a 240 V electric iron element if its current rating is 2.5 A.

18. How does the resistance of a conductor vary with (a) length, (b) cross-sectional area and (c) temperature? What else does the resistance of a conductor depend on?

19. (a) Define *resistivity*. (b) Calculate the resistance of a 66 m length of twin copper cable whose c.s.a. is 6 mm². ($\varrho = 1.7\ \mu\Omega$-cm).

20. A 50 m coil of tungsten wire has a c.s.a. of 0.125 mm² and a resistance of 22.5 Ω. Calculate the resistivity of tungsten.

21. What is the *temperature coefficient* of a conductor? Which common conducting material has a negative temperature coefficient?

22. A copper conductor has a resistance of 0.5 Ω when its temperature is 0°C. Calculate its resistance when the temperature is raised to 30°C ($\alpha = 0.004\ \Omega/\Omega/°C$).

23. With the aid of a diagram show how an ammeter and a voltmeter are connected in a circuit in order to measure the current taken by a resistive load and the voltage across it.

24. What types of supply are available and how are they produced?

25. What is the purpose of circuit protection? Where should protection be provided in a circuit?

Miscellaneous questions on Part I 153

VOLUME 1, CHAPTER 3

26. Two lamps of resistance 960 Ω and 1440 Ω respectively are connected in series. Calculate their combined resistance.

27. Three identical heating elements are connected in series and the total resistance is 172.8 Ω. What is the resistance of each element?

28. The same three elements (Question 27) are now connected in parallel. What is the total resistance?

29. An electric fire, a toaster, a kettle and a vacuum cleaner are all connected to a domestic supply. Their respective resistances are 19.2 Ω, 96 Ω, 28.8 Ω and 192 Ω. Calculate the total resistance of all these items.

30. Two heating elements of 28.8 Ω and 57.6 Ω are connected in series across a 240 V supply. Calculate the total resistance, total current, and voltage across each element.

31. Four lamps have the following resistances : 576 Ω, 960 Ω, 1440 Ω, 2304 Ω, and are connected across a 240 V supply. Calculate the total resistance, the total current, and the current through each lamp.

32. A cable of total resistance 0.2 Ω is used to supply a two-bar electric fire. Each element has a resistance of 576 Ω and they are connected in parallel. If the supply voltage is 240 V, calculate the volt drop along the cable.

33. Power in a circuit is given by $P = I \times V$. What two other formulae may be used?

34. Calculate the power dissipated by:

    (a) A 240 V fire taking 8.33 A.

    (b) An electric iron of resistance 96 Ω taking 2.5 A.

    (c) A 240 V vacuum cleaner having a resistance of 192 Ω.

35. What is meant by the term *rated value* when applied to an electrical appliance?

36. Calculate the power dissipated by a 100 W, 240 V lamp when connected to a 200 V supply.

154 *Miscellaneous questions on Part I*

37. A 1 kW, 240 V heating element is fed by a long cable of total resistance 2.4 Ω from a 240 V supply. Calculate the power dissipated by the heating element. Is the cable volt drop permissible?

38. With the aid of a diagram explain the action of a three-heat switch.

39. Draw a diagram of a light, controlled by a two-way and intermediate switching system. Show all the correct cable colours.

40. What is meant by (a) a final circuit, (b) a radial circuit and (c) a ring circuit?

41. What are the restrictions in the use of electrical apparatus in a room containing a fixed bath or shower?

42. What are the units in which electrical energy is measured? Calculate the energy consumed per week by a 300 W vacuum cleaner used for a total of 50 minutes each day.

43. A consumer has the following loads connected daily to a 240 V supply:

   Four 60 W lamps for 4 hours.

   One 600 W vacuum cleaner for 15 minutes.

   One 3 kW immersion heater for 2 hours.

   One 2 kW fan heater for 1.5 hours.

   One 3 kW kettle for 20 minutes.

   Sundry appliances take a total of 3 A for 1 hour. If the cost per unit is 2.7p, calculate his quarterly bill.

VOLUME 1, CHAPTER 4

44. State the three main effects of an electric current.

45. What is *corrosion*, and how can it be minimized in installation work?

46. A water heating system containing 60 litres of water at 15°C is supplied by a 2 kW element. How long would it take to heat the water to 51°C? (S.H. of water = 4180 J/kg/°C.) Ignore losses.

47. A water tank containing 240 litres of water has to be heated from 10°C to 80°C in three hours. If the efficiency of the tank is 92%,

calculate the nearest element size to do this. (S.H. of water = 4180 J/kg/°C.)

48. A conductor 15 cm long lying at right angles to a magnetic field of flux 15 mWb and c.s.a. 5 cm^2, carries a current of 6 A. Calculate the force on the conductor.

49. Show with the aid of a diagram how a conductor is caused to move in a magnetic field when a current is passed through the conductor.

50. What happens when a coil of wire is wound round (a) a hollow cylinder and a soft iron rod is placed inside, and (b) a soft iron core? How can these effects be used?

51. Explain with the aid of a diagram the action of a bimetal strip.

52. Show with sketches the action of (a) a thermal tripping mechanism, (b) a magnetic tripping mechanism and (c) a bellows-type air thermostat.

VOLUME 1, CHAPTER 5

53. Explain with the aid of a sketch the purpose of a simple commutator.

54. How is an alternating e.m.f. produced by a single loop of conductor rotating in a magnetic field?

55. With the aid of a sketch show how the basic components of a large single-phase a.c. generator are arranged in practice.

56. Sketch a typical output waveform for an a.c. generator. What is meant by (a) a cycle, (b) frequency and (c) peak value?

57. What is an *r.m.s.* value and how is it defined?

58. Draw the waveform over one complete cycle of a sinusoidal voltage having an r.m.s. value of 110 V. What is the value of the voltage after 60°?

59. Show how a three-phase a.c. supply is generated.

60. Sketch the output voltage waveform from a three-phase a.c. generator. What is the resultant value of the three voltages?

61. Why is it necessary to have a neutral conductor in a balanced three-phase distribution system?

62. Draw a diagram showing how a three-phase distribution system in a factory is balanced in practice.

VOLUME 1, CHAPTER 6

63. List the different methods used to generate electricity on a large scale. Which method is used most?

64. Explain briefly the steps taken to (a) dispose of waste and (b) control pollution, from power-stations.

65. List four advantages that exist because of the grid system.

66. State four advantages of the use of overhead lines to transmit electrical energy.

67. What are the main generation and transmission voltages used?

VOLUME 1, CHAPTER 7

68. With the aid of sketches, explain the difference between re-wirable, cartridge and H.R.C. fuses.

69. What is meant by *fusing factor?* A 30 A re-wirable fuse has a fusing factor of 1.7. At what value of current will the fuse operate?

70. For what purpose are H.R.C. fuses normally used and why?

71. What is meant by *discrimination?* Why is it important to arrange for discriminative protection?

72. State three advantages of the circuit breaker over the fuse.

73. With the aid of sketches explain the operation of a thermally operated indirect tripping mechanism.

74. What are the requirements of the I.E.E. Regulations with regard to: (i) the control of supply to a consumer's installation? (ii) the means of isolation? (iii) the sequence of control? and (iv) the connection of

control devices and switches?

75. The Regulations state that the current rating of every fuse fitted at the origin of a circuit must not exceed that of the lowest-rated conductor of the circuit. Explain the reasons for this.

VOLUME 1, CHAPTER 8

76. Explain the difference between cables used for fixed wiring, and flexible cords. What are the conductor colours for flexible cords?

77. What is *m.i.m.s. cable,* and what advantage does it have over P.V.C.-sheathed cable?

78. With the aid of a diagram show the 'screw-on pot' type of termination for mains cable.

79. Show with the aid of a sketch the construction of a P.V.C.-insulated, P.V.C.-sheathed steel wire armoured cable.

80. What are the requirements of the I.E.E. Regulations in respect of: (i) cables exposed to mechanical damage? (ii) cables passing through structural steelwork? and (iii) cables installed in a lift shaft?

81. Calculate the c.s.a. of a stranded conductor comprising seven strands each of 0.85 mm diameter.

82. What steps should be taken to ensure that all terminations are electrically and mechanically sound?

83. What precautions should be taken regarding P.V.C.-insulated cables in extremes of temperature?

84. Why must P.V.C.-insulated cables be kept clear of granular polystyrene loft insulation?

85. Why must the ends of steel conduit be reamed and bushed? Why and where should drain holes be provided in a finished conduit installation?

86. Explain the meaning of space factor.

158 *Miscellaneous questions on Part I*

87. What are the requirements of the I.E.E. Regulations with regard to: (i) bends in conduit? (ii) conduit installed in damp or corrosive situations? (iii) flexible metal conduit?

88. Why should all phase conductors and associated neutrals be contained in the same conduit or trunking?

89. What special precautions must be taken when installing rigid P.V.C. conduit in situations prone to extremes of temperature?

90. Explain with the aid of sketches how conduit may be terminated in metal trunking.

91. With the aid of a diagram explain the use of overhead bus-bar trunking.

92. What are the requirements of the I.E.E. Regulations in respect of trunking with regard to:

    (i) Preventing the spread of fire?

    (ii) Segregation of circuits?

    (iii) Expansion of conductors? (bus-bar trunking)

93. What special precautions must be taken in the use of (a) bolsters and cold chisels and (b) files?

94. List the tools required to install a conduit installation.

95. (a) How many teeth per 25 mm should a hacksaw blade have for use with conduit trunking?

    (b) What marking-out tools are required for the measurement and preparation of trunking for cutting?

96. What is the correct drilling tool for use with a hole saw? What other method can be used for cutting holes in trunking?

97. For what purpose are the following items used: (i) tallow? (ii) a scutch hammer? and (iii) stock?

98. What is (a) a thermoplastic material and (b) a thermosetting material?

99. What does 'P.V.C.' stand for and what type is usually used for cable insulation?

100. What is meant by the following terms: (i) hardening of metal? (ii) tempering of metal? (iii) annealing of metal? and (iv) cold working?

VOLUME 2, CHAPTER 1

101. (a) What is an *inductor?* Describe what happens when the d.c. supply to an inductor is switched off.

    (b) An electromagnet has 900 turns and when switched on the flux rises to 1.5 mWb in 300 ms. Calculate the value of the e.m.f. induced.

102. Define: (a) *self inductance;* (b) the *henry* and (c) *mutual inductance.*

103. Calculate the inductance of a 1000 turn coil which has a flux of 3 mWb when a current of 2 A flows.

104. With the aid of a sketch explain what happens to the current in a coil when a d.c. supply is applied. What is meant by *time constant?*

105. Calculate the energy stored in a coil of inductance 1.6 H when carrying a current of 0.4 A.

106. Explain what happens when an alternating current is applied to an inductor.

107. (a) What is *inductive reactance?* (b) An inductor has an inductance of 25 mH when connected to a 50 Hz supply. Calculate the inductive reactance of the coil.

108. An iron-cored coil has an inductance of 318 mH when connected to a 240 V, 50 Hz supply. Calculate the current taken.

109. Show with diagrams the waveforms and phasor diagrams of current and voltage for (a) a pure resistance and (b) a pure inductance.

110. A choke has a resistance of 80 Ω and an inductive reactance of 60 Ω. If the supply current is 2.4 A, calculate the voltage across the resistance and the inductance, and by phasor diagram determine the supply voltage.

111. What is *impedance?* Draw an impedance triangle.

160 *Miscellaneous questions on Part I*

112. A pure inductance of 30 Ω and a pure resistance of 40 Ω are connected in series across a 110 V supply. Calculate the circuit impedance and the circuit current.

113. Sketch the current phasor diagram for a pure inductance in parallel with a pure resistance.

114. Draw the power triangle of a circuit comprising a pure inductance in series with a pure resistance. Where is all the power dissipated to?

115. Define the term *power factor*. What is the practical importance of the power factor?

116. A single-phase load is rated at 5 kVA at 0.7 P.F. lagging. What is the kW rating of the load?

117. A single-phase motor gives the following instrument readings: 240 V; 0.5 A; 100 W. Calculate the P.F. of the motor.

118. How is an inductor utilized in a fluorescent lamp unit?

VOLUME 2, CHAPTER 2

119. What is a *capacitor?* Describe the construction of an electrolytic capacitor.

120. How does the capacitance of a capacitor vary with its dimensions?

121. (a) What is meant by the term *working voltage?* (b) Calculate the charge on a capacitor if its capacitance is 8 μF and the voltage applied is 500 V.

122. Calculate the energy stored in a capacitor of 20 μF if it is charged at 240 V.

123. Sketch the curve of the current decay that occurs when a capacitor is discharged.

124. The following capacitors are connected first in series then in parallel: 80 μF; 120 μF; and 52 μF. If the supply voltage is 240 V, calculate the total charge in each case.

125. A parallel-plate capacitor has a reactance of 15.9 Ω when connected

to a 50 Hz supply. Calculate its capacitance.

126. A capacitor of reactance 20 Ω and a pure inductor of reactance 10 Ω are connected in series across a 240 V supply. Calculate the current taken from the supply.

127. A capacitor of value 8 μF is connected in parallel with a resistance of 200 Ω across a 240 V supply. Calculate the current taken by each component, and by means of a phasor diagram, determine the value of the supply current.

VOLUME 2, CHAPTER 3

128. With the aid of diagrams, show how the power factor of an inductive load such as a fluorescent lighting unit may be improved.

129. Why is it desirable to improve the power factor? Why is P.F. improved to a value of less than unity?

130. What methods may be used to improve the power factor?

131. A motor takes a current of 10 A at a P.F. of 0.78 lagging. Determine the current taken by the supply when the P.F. is improved to 0.93 lagging.

132. Draw the circuit diagram of a single fluorescent lighting unit. Explain how it operates.

133. A 240 V, 50 Hz single fluorescent luminaire, without a P.F. correction capacitor, takes a current of 0.8 A at 0.58 lagging. With the aid of a phasor diagram determine the value of the capacitor current needed to improve the P.F. to 0.96 lagging. Use this value to calculate the value of the capacitor.

134. What are the requirements of the I.E.E. Regulations in respect of: (i) current rating of fluorescent lighting circuits? and (ii) the rating and type of switch used to control inductive circuits.

135. Calculate the number of 125 W, 240 V fluorescent luminaires that may be controlled by an ordinary 5 A switch.

136. What is meant by the *stroboscopic effect* of a fluorescent lamp? How may this effect be overcome?

162 *Miscellaneous questions on Part I*

137. Draw the circuit diagram of a lead-lag circuit for fluorescent luminaires.

VOLUME 2, CHAPTER 4

138. What is a *primary cell?* Describe the construction of a dry cell.

139. What is a *secondary cell?* Describe the construction of a lead-acid cell.

140. Describe in detail a routine weekly maintenance procedure carried out on a lead-acid cell. Include all the values that are expected.

141. Describe the construction of an alkaline cell. What kind of maintenance is required?

142. What is meant by the following terms: (i) *capacity* of a cell? and (ii) *ampere-hour efficiency* of a cell?

143. Sketch graphs comparing the charge and discharge characteristics of lead-acid and alkaline cells.

144. List the advantages and disadvantages of lead-acid and alkaline cells.

145. What is the difference between the e.m.f. of a cell and its terminal voltage?

146. A cell having an e.m.f. of 2 V and an internal resistance of $0.16\,\Omega$ is connected across a load of resistance $15\,\Omega$. Calculate the current taken from the cell and its terminal voltage.

147. Explain what advantages may be gained by connecting cells in (a) series and (b) parallel.

148. State two popular methods of battery charging. Illustrate one of these by means of a circuit diagram.

149. Explain with the aid of a diagram the action of a simple transformer.

150. A 240 V/8 V bell transformer has 720 turns on its primary winding. How many turns are there on the secondary winding?

151. What is an *auto-transformer?* Why is the step-down type not recommended in domestic situations? Why are they used as Grid transformers?

152. With the aid of a sketch explain the construction and purpose of a bar primary current transformer.

153. List the losses which occur in transformers.

154. What is *hysteresis?* How does it affect transformers?

155. How can eddy current and hysteresis losses be minimized?

156. What are the recommendations of the I.E.E. Regulations with regard to hand-held appliances in damp situations?

157. With the aid of a diagram explain the action of (a) a closed-circuit and (b) an open-circuit alarm system.

158. Give the advantages and disadvantages of open- and closed-circuit burglar alarm systems.

159. Draw the circuit diagram of a simple four-button call/indicator system.

160. What are the two main types of emergency lighting system? Draw a simple block diagram of one of these.

VOLUME 2, CHAPTER 5

161. What is meant by *diversity?* What advantage is gained by applying it?

162. A house has the following loads:

    Ten lighting points
    Two 30 A ring final circuits
    One 20 A radial circuit
    One 9.6 kW cooker (unit has socket outlet)
    One 2 kW thermostatically controlled water heater.

    Calculate the minimum size of meter tails if the supply is 240 V and the protection is by type 2 M.C.B.'s.

164   *Miscellaneous questions on Part I*

163. What is the purpose of the factors shown in the I.E.E. Regulations?

164. A 10 kW load is to be supplied at 240 V by a two-core P.V.C.-armoured copper cable, 25 m long, which is clipped to a cable tray together with four other cables. If the ambient temperature is 50°C and the protection is afforded by a circuit breaker, calculate the minimum size of cable permissible. What is the volt drop along the cable?

165. A 240 V final circuit cable of 6.0 mm² twin P.V.C. copper is 21 m long and is run clipped to a surface in an ambient temperature of 45°C. If the volt drop between the supply fuse and the distribution board is 1.5 V, calculate the maximum load that may be supplied (protection is by re-wirable fuse).

VOLUME 2, CHAPTER 6

166. Draw a block diagram showing the sequence of control of a three-phase supply in a consumer's premises.

167. What are the I.E.E. Regulations regarding the methods used to prevent dangerous earth-leakage currents?

168. With the aid of a diagram show what is meant by the *earth-fault loop path*. List its component parts.

169. Explain why the value of the earth loop impedance is so important.

170. Give an example of equipment protected by 'basic insulation'.

171. Suggest why the general mass of earth as part of the earth path could be unreliable.

172. When is it necessary to install an earth leakage circuit breaker?

173. Draw a connection diagram to show how and where an R.C.D. may be installed.

174. With the aid of diagrams explain the operation of an R.C.D.

175. What are the requirements of the I.E.E. Regulations in respect of: (i) bonding of services; (ii) bonding of metalwork?

VOLUME 2, CHAPTER 7

176. List, in sequence, the tests to be made on a completed installation, as required by the I.E.E. Regulations.

177. What is the purpose of a polarity test? Describe a method of making such a test. What test instruments would be used?

178. Why is it necessary to test the circuit protective conductor? What tests may be carried out? Detail one of these tests. What values would you expect?

179. Explain the difference between a phase-earth and a neutral-earth loop impedance test. Include diagrams.

180. Describe the tests to be carried out for insulation resistance on (a) circuit wiring, and (b) appliances. What values would you hope to measure?

181. Explain why the insulation resistances of separate circuits must be added in parallel to determine the overall resistance of an installation.

182. The following readings were obtained when an insulation resistance test was carried out on a consumer's installation:

    upstairs lighting — 60 MΩ
    upstairs ring main — 100 MΩ
    downstairs lighting — 250 MΩ
    downstairs ring main — 500 MΩ
    circuit for cooker — 1.5 MΩ
    circuit for water heater — 3 MΩ

    What would have been the reading, taken with all fuses in place, at the tails of the distribution board?

183. Explain with the aid of a diagram the test for ring continuity.

184. You are required to make a visual inspection of a consumer's premises. List the points you would look for.

185. Explain with the aid of a diagram the operation of a moving-iron repulsion instrument.

186. Explain with the aid of a diagram the operation of a moving-coil instrument.

187. Compare the advantages and disadvantages of moving-coil and moving-iron instruments.

VOLUME 2, CHAPTER 8

188. Write down the value of the resistors with the following colour codes:

   (a) Orange-silver-red-silver
   (b) Red-white-green-red
   (c) Green-green-black-gold-brown
   (d) Brown-red-gold-silver.

189. Give the colour code for the following values of resistor:

   (a) 8.7K 1%
   (b) 15K8J
   (c) 100RG
   (d) 350K 20%.

190. Indicate the values of the following colour coded capacitors:

   (a) Brown-silver-gold-green-brown
   (b) Red-violet-gold-white-yellow
   (c) Violet-green-gold-black-brown
   (d) Brown-black-yellow-white-red.

191. Write down the colour code for the following capacitors:

   (a) 25 nF 20% 400 V
   (b) 53 nF 10% 100 V
   (c) 15 nF 5% 250V
   (d) 200 nF 20% 250 V.

192. Explain with the aid of diagrams the difference between a heat sink and a heat shunt.

193. Draw the B.S. 3939 symbols for the following components:

   (a) an L.E.D.
   (b) a PNP transistor
   (c) an iron-cored inductor
   (d) a variable capacitor.

194. With the aid of drawings explain the difference between block, circuit and layout diagrams.

Answers to Test Questions

Answers to Test Questions

Chapter 1
1   30 turns
2   1 henry
3   0.08 s; 20 A; 2 A
4 (b)   5 A
5   50 Hz
6   240 V
7   240 V
8   20 A; 30 A; 36.05 A
9(b)   240 V; 144 V
10   0.6

Chapter 2
1   7.3 mC
2   1.92 mC
3   10 μF
4   50 μF; 12 mC; 4.8 mC
5   80 cm^2
6   50 Hz
7   144 V; 192 V; 240 V
9   9.6 A; 4.8 A; 10.7 A

Chapter 3
1   6.25 A; 82.8 μF
2   1.27 A; 0.94 lagging

Chapter 4
4(b)   0.516 Ω; 1.55 V
6(b)   330

Answers to Miscellaneous Questions on Part I

5	588.6 N	108	2.4 A
16	3 A	110	240 V
17	96 Ω	112	50 Ω; 2.2 A
19	0.374 Ω	116	3.5 kW
20	56.2 $\mu\Omega$-mm	117	0.833 lagging
22	0.56 Ω	121	4 mC
26	2440 Ω	122	576 mJ
27	57.6 Ω	124	6 mC; 60.5 mC
28	19.2 Ω	125	200 μF
29	9.76 Ω	126	24 A
30	86.4 Ω; 2.77 A; 80 V; 160 V	127	0.6 A; 1.2 A; 1.34 A
31	256 Ω; 0.94 A; 0.4166 A; 0.25 A; 0.166 A; 0.104 A	131	8.4 A
		133	0.518A; 6.8 μF
32	0.166 V	135	2
34	(a) 2 kW (b) 600 W (c) 300 W	146	132 mA; 1.98 V
36	69.4 W	150	24 turns
37	921.6 W; 9.6 V not acceptable	162	16 mm^2
42	1.75 kWh	164	25 mm^2; 1.77 V
43	£29.06	165	7.56 kW
46	1.25 hours	171	1.67 Ω
47	7 kW	182	0.968 MΩ
48	27 N	188	(a) 3.8 k 10% (b) 29.5 k 5% (c) 550 k 1% (d) 12 k 10%
58	134.7 V		
69	51 A	189	(a) Silver-violet-red-brown (b) Brown-green-silver-red-green (c) Brown-black-black-silver (d) Gold-green-black-gold
70	4.0 mm^2		
101	-4.5 V		
103	1.5 H		
105	128 mJ	190	(a) 18 nF 5% 100 V (b) 27 nF 10% 400 V
107	7.85 Ω		

Answers to Miscellaneous Questions on Part 1

(c) 75 nF 20% 100 V
(d) 100 nF 10% 250 V

191 (a) Brown-green-gold-black-yellow
(b) Green-red-gold-white-brown
(c) Brown-green-gold-green-red
(d) Red-black-yellow-black-red

Index

All-insulated construction 110
Argon gas 37
Atoms (gas) 37

Back e.m.f. 1
Bell and call system 64
Bimetallic-type starter contacts 38
Burglar alarm 62
   open-circuit 65
   closed-circuit 66
British Standard 3939 76, 77ff, 97

Call point 65
Call system 64
Capacitance 22, 26
   in a.c. circuits 28
Capacitive reactance 28
Capacitor,
   charging 26
   dimensions 23
   discharging 27
   electrolytic 22
   energy stored in 26
   in electronics 139
   in parallel 25
   in series 23
   time constant 27
   working voltage 31
Capacitance and resistance in parallel 30
Capacitance and resistance in series 29
Capacitor as suppressor 38
Cells and batteries,
   alkaline 50
   capacity of 51
   characteristics 51
   charging methods 55
   efficiency 51
   internal e.m.f. 52
   internal resistance 52
   lead-acid 47

   maintenance 50, 51
   primary 45
   secondary 46
   specific gravity 48
   terminal voltage 53
Charge 22
Choke 38
Circuit breaker 115
Circuit diagram, as distinct from wiring diagram 63
Coercive force 60
Colour coding 138, 140
Consumer unit 111
Control 111
Copper loss 60
Correction factor 106

Damp and corrosive situations 108
Definitions of some terms used in I.E.E. Regulations 87ff
Depolarizer 45
Diac 142
Dielectric 22
Diversity 100
Double insulation 113

Earth as a conductor 114
Earth fault loop path 114
Earth leakage 115
Earth leakage circuit breakers, residual current-operated 115
Earth loop impedance 114
Earthing, of exposed metalwork 113
Earthing clamp 117
Earthing conductor 113
Earthing terminal 117
Eddy currents 60
Electrolyte 45
Electrons 60
Element of fluorescent lamp 37
Emergency lighting 68

Index

Energy stored in capacitor 26
Energy stored in magnetic field 7
Estimates 99
Explosive situations 107
Extra-low voltage 62

Farad 22
Fire alarm, *see* Burglar alarm
Flammable situations 107
Fluorescent lighting 37
    rating of 40
    stroboscopic effect 41
Flux 1

Heat shunt 150
Heat sink 142
Henry 3
'Hold-on' circuit 63
Hydrometer 48
Hysteresis 60

Impedance 12
Inductance 1
    mutual 3
    self 3
Induced e.m.f. 2
Inductive reactance 8
Inductors in electronics 140
Inspection 129
Instruments,
    insulation resistance 135
    phase-earth loop tester 136
    moving-coil 133
    moving-iron 132
    ohmmeter 134
Ionization 37
Isolation 110
Isolator 110

Junction diode 141

Lamination 60
Lead-lag circuit 42
L.E.D. 142

Loop impedance test 123
L.S.D. 142

Magnetizing force 60
Materials list 104
Mercury vapour 37
Mutual inductance 3

Oxide coating on fluorescent light
    element 38

Phasors 10
Power, in a.c. circuits 17
Power factor 18
    improvement of 33

Rectifier 61
Relay 63
Requisitions 99
Residual magnetism 60
Resistance and inductance in series 11
Resistors 138
RKM coding 138

Safety and welfare 76
Saturation 60
Schematic diagram, as distinct
    from wiring diagram 63
Security 103
Semi-conductors 140
Sequence of control 112
Shaver unit 62
Soldering 149
Specification 70
Starter (fluorescent) 38
Stroboscopic effect 41
Symbols for electronics components
    144

Terminal voltage 53
Testing 119
Thermistor 142
Thyristor 141
Time constant 4, 27

Transformer 56
   auto 58
   current 59
   double-wound 56
   in electronics 140
   losses from 59
   portable tools, use with 62
Transformer ratios 57
Transistor 141

Triac 142

Ultra-violet light 38

Wiring diagram, as distinct from circuit diagram 63
Working voltage 31

Zener diode 141